NASA/ESA/ASI
CASSINI-HUYGENS
1997–2017
(Cassini orbiter, Huygens probe and future exploration concepts)

First published in April 2017

A catalogue record for this book is available from the British Library.

ISBN 978 1 78521 111 9

Library of Congress control no. 2016959355

Published by Haynes Publishing,
Sparkford, Yeovil,
Somerset BA22 7JJ, UK.
Tel: 01963 440635
Int. tel: +44 1963 440635
Website: www.haynes.com

Haynes North America Inc.,
861 Lawrence Drive, Newbury Park,
California 91320, USA.

Printed in Malaysia.

NASA/ESA/ASI CASSINI-HUYGENS

1997–2017

(Cassini orbiter, Huygens probe and future exploration concepts)

Owners' Workshop Manual

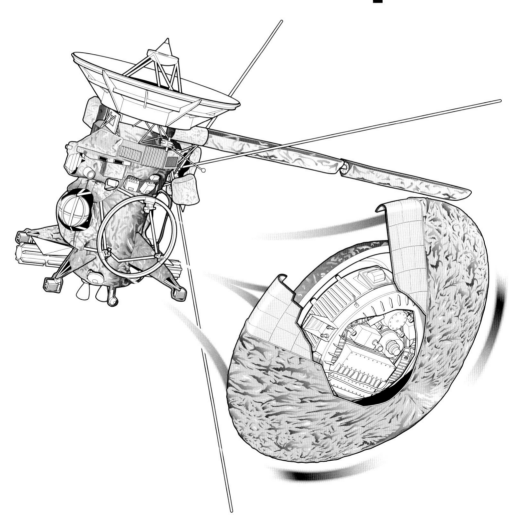

An insight into the technology, mission planning and operation of spacecraft designed to study Saturn's moon, Titan and the Saturnian System

Ralph Lorenz

Contents

OPPOSITE **Artist's impression circa 1988 of Cassini with its (as yet-unnamed) Titan probe. The Cassini orbiter spacecraft portrayed here has a separate probe relay antenna, teardrop fuel tanks, a single main rocket engine and scan platforms for instruments. All these details changed as the mission evolved towards launch in 1997.** *(NASA)*

Preface

The book is intended as a paean to the project on which – and the remarkable people with whom – I have spent my entire career at four institutions in three countries over nearly three decades. While the scientific findings of the Cassini-Huygens mission are covered in several good popular accounts, and many technical papers exist on specific topics, a broad and accessible yet detailed treatment of the hardware and how it was assembled, tested, and operated was lacking. The end of the Cassini mission in 2017 is an appropriate time to develop such a synthesis, and I was therefore thrilled that Haynes was willing to publish an Owners' Workshop Manual on this topic.

I wanted to remain faithful to the ethos of the original Haynes Manuals for car repair in portraying the mechanical assembly in rich graphic detail, but of course the number of parts and functions on Cassini-Huygens is far too large to cover completely, and much of the detail would be tedious, proprietary and/ or subject to US export control. However, I hope I have succeeded in showing details of at least a few branches of the tree, to hint at the nested levels of artful design, meticulous craftsmanship, and diligent operation that pervade the whole project. While some graphics may be familiar to close followers of Cassini-Huygens, I have striven to show

BELOW **The author with the Huygens probe (the SM2 parachute drop-test model) at the European Space Operations Center in Darmstadt, Germany in January 2005.** *(Author)*

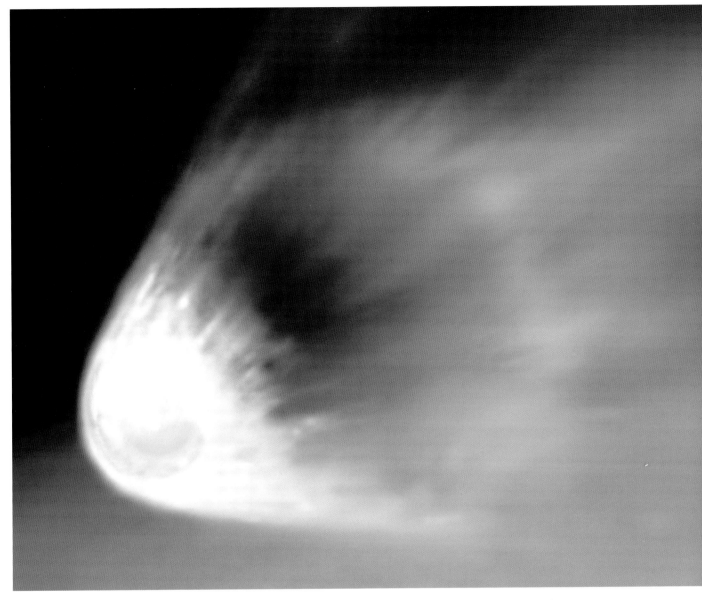

material not widely seen, portraying the nuts and bolts of spacecraft and instrument design and operation, the practical details that are not typically covered in aerospace engineering textbooks – including aspects of the mission that have not gone as planned. I hope in this respect the book may be useful to future space pioneers.

It is reckoned that around 5,000 people have had some kind of close involvement in Cassini-Huygens's development and flight. There is no room to name everyone, so I have studiously avoided naming anyone in the text, except Messrs Huygens and Cassini themselves!

The rewards for participating in an enterprise on the scale of Cassini-Huygens, however,

come not from being named in books, or receiving honours and certificates, but from the act itself. In an international effort that spans decades and stretches across a billion miles of space, the development of these amazing machines to explore distant worlds represents a triumph of human ingenuity, and a sustained unity of noble purpose that transcends borders, disciplines and generations. Planetary exploration is simply one of the best endeavours that human beings can ever hope to pursue. It is, then, a great privilege for me to have been a small part of the Cassini-Huygens adventure, and to tell at least a part of its story.

Ralph Lorenz, November 2016

ABOVE Designers of planetary probes must anticipate the diverse challenges of alien environments – the vacuum of space, the dense, frigid atmosphere of Titan, and shown here, the brutal aerodynamic and aerothermodynamic loads of hypersonic entry. *(J. Garry)*

Chapter One

Origins of Cassini

The technical design of Cassini-Huygens owes heritage to the Pioneer, Voyager and spacecraft that preceded it into space, but the mysteries Cassini was built to solve date back to the first days of telescope astronomy and the two scientists after which the mission is named.

OPPOSITE Pioneer 10 and 11, the first explorers to the outer solar system, set the pattern for the design of spacecraft far from the Sun and Earth. The spacecraft is dominated by a large dish antenna for communications, and two finned Radioisotope Thermoelectric Generators (at top and lower right) for electrical power. At left is a magnetometer instrument mounted on a boom to minimise disturbance fields from the spacecraft itself. *(NASA)*

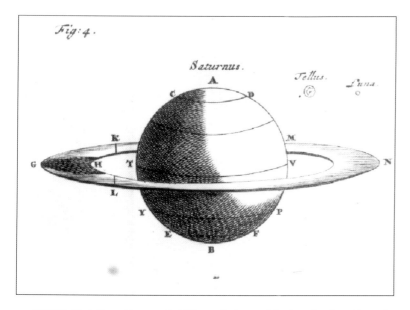

Exploration of the Outer Solar System

The planet Saturn, the most distant planet easily visible to the naked eye, has been known since ancient times. But the invention of the first blurry telescopes in the early 17th century revealed it to be special – there was something more than just an orb. The Dutch astronomer, Christiaan Huygens, with a better telescope in the mid-1650s discerned the remarkable feature of this world: it possessed a ring. Huygens also found that Saturn had a giant moon that would later be named Titan, and that Titan, the rings, and Saturn's equator were all inclined to Saturn's orbit around the Sun so that this system has seasons like those of Earth.

But being so far from the Sun (ten times further than Earth) these seasons would be long because Saturn takes nearly 30 Earth years to orbit the Sun. And, Huygens realised, Saturn must be cold, meaning that if there was rain out there it would have to be rain of something other than water.

Among Huygens's contemporaries was an Italian-born astronomer, Giovanni Domenico Cassini, who established the Paris Observatory and became better known in his adopted country as Jean-Dominique Cassini. He found that Saturn's rings were not continuous, but had a dark gap, now named the Cassini Division. He also found several other moons.

The Saturnian system was full of mysteries, but Huygens calculated that if a bullet, the fastest thing he knew about, travelled 100

ABOVE Christiaan Huygens's billion-mile leap of the imagination. Not only did Huygens discover Titan and ascertain correctly the architecture of the rings, he expressed his knowledge in a picture which showed a view that cannot be obtained from Earth. He mentally transported himself out to Saturn to visualise how in summer the planet would cast a curved shadow on its rings, but with the shadow generally hidden from our view. He wrote in Latin: Tellus and Luna show the size of the Earth and Moon relative to Saturn. *(C. Huygens)*

BELOW Seasons on another planet. This admirable graphic from Christiaan Huygens's writings shows not only Saturn's relationship to the Sun and Earth, but also how its appearance from Earth (outer ring of 'simulated' views) varies due to the fixed orientation of Saturn's ring and pole with respect to space (the inner ring of Saturn sketches). The Sun is in the centre, and the Earth's position is shown. The Zodiacal directions in the sky are also noted – at about one o'clock is the First Point of Aries, indicated by a curly 'v' to suggest a ram, which is a common astronomical reference direction. *(C. Huygens)*

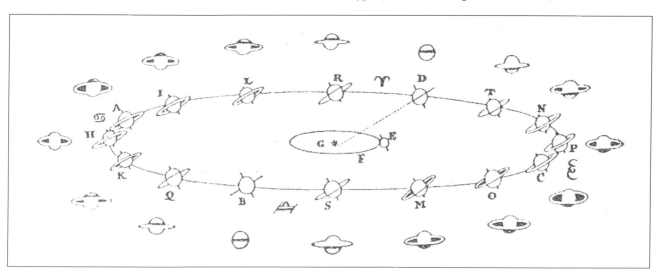

fathoms in a heartbeat, it would take 250 years to reach Saturn. Clearly, the mysteries would not be resolved for centuries.

The names of these two astronomers have been given to the most ambitious international undertaking in planetary exploration, a pair of spacecraft which, between them, have transformed our understanding of the Saturnian system. The Cassini-Huygens mission was an enterprise spanning more than a quarter of a century, the product of billions of dollars of investment, amazing technological advances, and the efforts of thousands of engineers and scientists on two continents.

Cassini discovered four moons of Saturn, including Iapetus, which appeared to have a dark hemisphere facing the direction that it travelled around the planet. The following centuries saw another three moons racked up to astronomers looking through telescopes. Then in 1899, using photographic plates, a small outer satellite, Phoebe, was discovered.

An astronomer's eyeball drew – literally, the observation was recorded as a sketch – further attention to the Saturnian system in 1907, when an astronomer in Barcelona discerned that Titan's tiny disk did not have a hard edge like the Moon, but was softly dimmed, suggesting that unlike any other satellite in the solar system, it possessed an atmosphere.

This finding was confirmed in 1944, when spectroscopic observations showed that indeed sunlight was not reflected from Titan equally at all wavelengths – in particular, specific bands

of red and infrared light were absorbed by the moon: the spectral fingerprint of methane gas. Then, in the 1970s, as planetary exploration by robotic spacecraft began in earnest, it was realised that Titan's atmosphere was hazy, and that sunlight would slowly destroy the methane, forming other, more complex, organic molecules. It was speculated that this process may have happened on the early Earth, and might be the first stepping stone in chemical evolution that produces the building blocks of life. Titan, then, was set as a priority target for exploration.

A lightly instrumented pathfinder spacecraft, Pioneer 11, made a brief reconnaissance of the Saturnian system in 1979, measuring radiation levels and assessing the hazard of impacts with particles near the rings. Designed at NASA's Ames Research Center near San

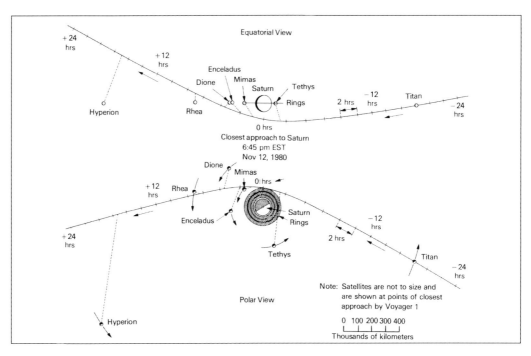

Francisco, this spacecraft did not have a camera, but its photopolarimeter, a light sensor in a tube that was swept around as the craft spun on its axis, did permit the construction of some rudimentary images.

But it was Voyager 1 in 1980, and its companion a year later (both built at the Jet Propulsion Laboratory near Los Angeles), that really unveiled the Saturn system. Its vidicon camera tubes showed bizarre details in the rings, including kinks and corrugations, and documented the surfaces of Saturn's airless moons as a diverse menagerie. There was Mimas, with a giant crater making it look like the 'Death Star' from *Star Wars* (which had been released three years earlier); yin-yang Iapetus, whose pitch-black terrain formed an ellipsoidal patch like half a tennis ball on one hemisphere; Tethys, spanned by a giant canyon where its crust seemed to have stretched open; and snow-white Enceladus with regions strangely smooth and free of craters. Titan's surface was shrouded in orange smog, but Voyager's infrared spectrometer detected a dozen organic compounds, such as propane, acetylene and cyanogen, as well as carbon dioxide, and the ultraviolet spectrometer showed that the atmosphere, like our own, had molecular nitrogen.

But how thick was it? Voyager 1 was targeted to fly behind Titan as seen from Earth, and to send a radio signal through the atmosphere. By measuring how much the atmosphere bent the radio rays (much as if it was a lens), scientists could calculate the density of the gas. This radio occultation technique showed that Titan's atmosphere was four times denser than air at sea level on Earth, and that, despite the greenhouse effect of methane, it was bitterly cold at 94K (minus 179°C).

These conditions were disappointing to hopes for a hospitably warm surface, but nonetheless posed the interesting possibility that methane might exist as a liquid on Titan's surface. But before the Voyagers – or even Pioneer 11 – reached Saturn, scientists were thinking ahead. A logical next step to explore Saturn and Titan would be something like the Galileo mission being developed for Jupiter that featured an orbiter to survey the planet, its satellites and magnetosphere for an extended period, plus a detachable entry probe to investigate the planet's atmosphere in detail.

In 1976 NASA had contracted the company Martin Marietta to study Titan exploration ideas, and many interesting possibilities were identified such as balloons (see Chapter 9). But a major challenge was the uncertainty in the thickness of Titan's atmosphere: it could perhaps be just a bit thicker than that of Mars (Marietta had built the Viking landers which made the first successful soft-landings on that planet that same year), or perhaps much thicker than Earth's. The most efficient designs would

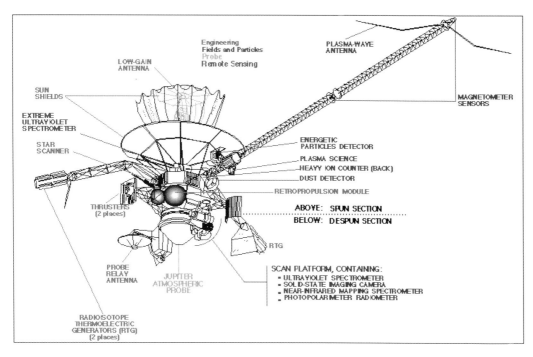

LOW-GAIN ANTENNA

Engineering
Fields and Particles
Probe
Remote Sensing

PLASMA-WAVE ANTENNA

SUN SHIELDS

EXTREME ULTRAVIOLET SPECTROMETER

STAR SCANNER

MAGNETOMETER SENSORS

ENERGETIC PARTICLES DETECTOR
PLASMA SCIENCE
HEAVY ION COUNTER (BACK)
DUST DETECTOR

RETROPROPULSION MODULE

THRUSTERS (2 places)

ABOVE: SPUN SECTION
BELOW: DESPUN SECTION

RTG

PROBE RELAY ANTENNA

JUPITER ATMOSPHERIC PROBE

SCAN PLATFORM, CONTAINING:
▪ ULTRAVIOLET SPECTROMETER
▪ SOLID-STATE IMAGING CAMERA
▪ NEAR-INFRARED MAPPING SPECTROMETER
▪ PHOTOPOLARIMETER RADIOMETER

RADIOISOTOPE THERMOELECTRIC GENERATORS (RTG) (2 places)

LEFT Sketch of the Galileo orbiter with the partly deployed high gain reflector. The main antenna, RTGs and particles and fields instruments are on a spun section that rotated for stability and for instrument scanning; and the probe, probe relay antenna, and scan platform carrying the optical remote sensing instruments are on the non-rotating (despun) section. *(NASA/JPL)*

depend on knowing how dense the atmosphere would be.

In 1977–78 NASA began studying a Saturn Orbiter/Dual Probe (or 'SO2P') concept, with a probe for Titan and a probe for Saturn. NASA's Ames Research Center, which had developed the four probes for the Pioneer Venus mission that operated there in 1979, was to study the probes while JPL managed the study overall – roles similar to those these organisations would pursue on the Galileo mission. (Galileo, whose propulsion system was supplied by Germany, was ready to fly in 1986 but the disastrous loss of the Challenger Space Shuttle in January 1986 would delay its launch by several years.)

Across the Atlantic, Europe's own capabilities in solar system exploration were emerging. A joint European–NASA International Solar Polar Mission (ISPM) was planned in 1977. Ironically, the two spacecraft would fly out to Jupiter to use its gravity to depart the ecliptic plane in which most planets travel

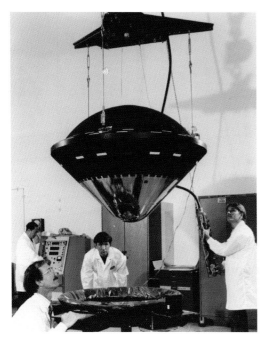

FAR LEFT The Galileo probe, encapsulated in its massive heatshield, here seen at NASA's Ames Research Center. *(NASA)*

LEFT The Galileo probe suspended above its heatshield, the back side of which is visible at the bottom. Various instrument ports are visible on the probe's hull, in some cases with covers. *(NASA)*

TITAN RADAR MAPPER

- SURFACE IMAGING
- ALTIMETRY
- SUB-SURFACE SOUNDING

ABOVE The original concept intended a radar instrument that could image strips on both sides, as well as measuring the height (altimetry) and subsurface structure (sounding) along the ground-track. *(NASA/JPL)*

RIGHT ESA's Giotto spacecraft with a cylindrical solar array, dustshield at the bottom, and despun dish antenna at the top. Its science instruments included a camera (the white cylinder at the lower centre), dust impact sensors on the shield, and a magnetometer at the apex of the antenna tripod. Many European scientists associated with these experiments would later work on Cassini-Huygens. *(ESA)*

around the Sun, therefore they would need Radioisotope Thermoelectric Generators like those of Voyager.

The Voyager findings laid the intellectual and technical stepping stones for the mission to follow. They identified many compelling scientific questions about the origin of Saturn's diverse satellites and exotic ring system, as well as one overpowering and succinct mystery: What was the surface of Titan like? The atmospheric profile and composition gave the tantalising

prospect that liquid methane might exist on the surface, and the tools to solve the mystery – a descent probe and the use of a mapping radar – were clear and ready (at this point, after some crude radar observations by Pioneer Venus, a Venus radar mapping mission, later to be named Magellan, was under development, and the analogy of exploring our cloud-shrouded sister planet by radar and doing the same at Titan was obvious).

Halley's comet was on its way to the inner solar system and an international armada of spacecraft would fly out to greet it, but NASA's plans for a dedicated Halley spacecraft were sadly cancelled. In 1980 the European Space Agency began development of Giotto, a small plucky spin-stabilised spacecraft that, in 1986, would fly closer to the comet than the two Japanese and two Soviet missions.

In 1981, NASA's Space Shuttle made its first flight and the Cold War was at its height. NASA's budget was being squeezed and focused upon human spaceflight, and the US involvement in ISPM was scaled back. Now there would be only a single (European) spacecraft, albeit with the significant contribution of an American Radioisotope Thermoelectric Generator and a launch on the Space Shuttle. The spacecraft, to be called Ulysses, would carry instruments from both the USA and Europe, and communications would be provided by NASA's Deep Space Network. Ulysses was originally to launch in May 1986, but was delayed by the Challenger disaster. On being released from the cargo bay of the Discovery in 1990, the spacecraft was accelerated out of Earth orbit on a large three-stage solid rocket booster.

Despite the challenges of cooperation with the USA, when ESA issued a call for ideas for future missions a proposal was made in 1982 for a mission to Saturn and Titan, to be called 'Cassini'. It was imagined that this would have a European spacecraft, perhaps patterned on Giotto, and a NASA probe similar to that of Galileo. The proposers stressed the wide range of scientific disciplines that could be addressed by such a mission, ranging from the astrodynamics of the rings, to the geology of the satellites, to chemistry in the atmospheres and the electrodynamics of the plasma in the magnetosphere. Such a broad scientific appeal generated strong political support, and based on this idea the mission took shape with various iterations of committees in Europe and the USA.

In the 1980s JPL advocated a common large spacecraft 'bus' to provide the basic functions for a variety of outer solar system missions, called 'Mariner Mark II', harking back to the original Mariner series which visited Mars, Venus and Mercury, with modest adaptations for the different destinations. The structures, propulsion systems and so on would be designed only once for general use, then the instruments and any sub-vehicles would be tailored to the specific application.

Two initial missions were considered: Cassini and a Comet Rendezvous-Asteroid Flyby (CRAF). Both would require large instrument suites mounted on scan platforms that could point in different directions independently of the main spacecraft, on which there would be a large dish antenna for communications. Cassini would carry a Titan Atmosphere Probe, and CRAF would carry a slender 'penetrator' that would be fired into the nucleus of a comet. The roles of Europe and the USA were reversed from the original proposal – with NASA providing the 'mother ship' and ESA providing the probe. This would be a new (and therefore technologically appealing) endeavour for the European aerospace industry, requiring systems such as a heatshield and parachutes not normally found on satellites.

The earliest concept of a Galileo-like spacecraft (an idea later resurrected in response to budget pressures) had some technical difficulties. First, Cassini's more ambitious payload, notably including a radar, would be difficult to accommodate on the dual-spin spacecraft. Second, the Galileo probe was mounted in front of the spacecraft's main engine, which was fine for a probe to Jupiter that was to be released six months in advance of arrival. For a Titan probe, the engine would be needed to brake into orbit around Saturn first, and so the probe would be in the way. Major redesign would be needed, whatever happened.

Initially it was imagined that like Ulysses and Galileo, Cassini would be deployed by the Space Shuttle. However, Cassini would be much larger than these spacecraft, and had

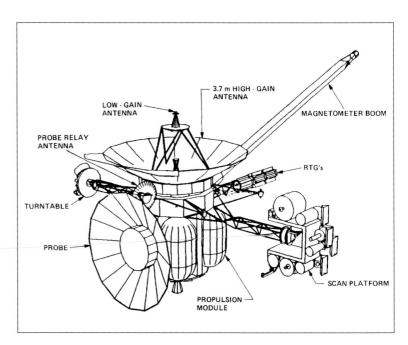

ABOVE The original concept for Cassini envisaged it being launched on the Space Shuttle. Here egg-shaped propellant tanks are shown, and a ponderous scan platform. Note that the probe was to be mounted nose-inwards, to minimise the structural load path to the most massive part. *(NASA/JPL)*

BELOW Impression of ESA's spin-stabilised Ulysses spacecraft after release from the Space Shuttle. The spacecraft at the left is dwarfed by its three-stage solid rocket motor. Note the lattice to support the cantilevered finned Radioisotope Thermoelectric Generator against the acceleration of the solid motor. *(NASA)*

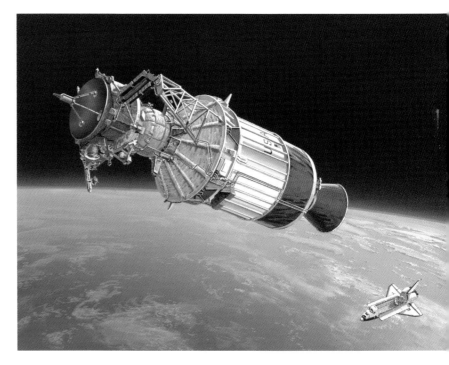

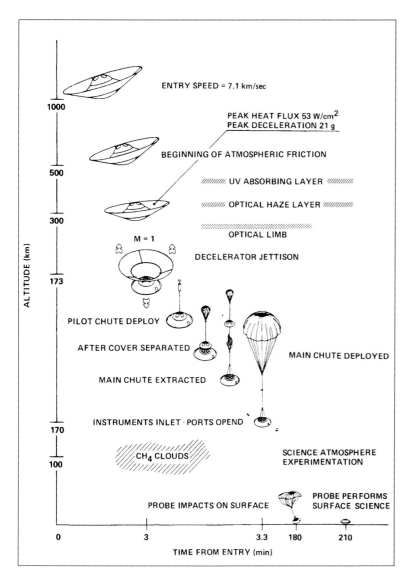

ENTRY SPEED = 7.1 km/sec

PEAK HEAT FLUX 53 W/cm^2
PEAK DECELERATION 21 g

BEGINNING OF ATMOSPHERIC FRICTION

UV ABSORBING LAYER

OPTICAL HAZE LAYER

OPTICAL LIMB

M = 1

DECELERATOR JETTISON

PILOT CHUTE DEPLOY

AFTER COVER SEPARATED

MAIN CHUTE DEPLOYED

MAIN CHUTE EXTRACTED

INSTRUMENTS INLET · PORTS OPEND

CH$_4$ CLOUDS

SCIENCE ATMOSPHERE
EXPERIMENTATION

PROBE IMPACTS ON SURFACE

PROBE PERFORMS
SURFACE SCIENCE

ALTITUDE (km)

1000
500
300
173
170
100

TIME FROM ENTRY (min)
0   3   3.3   180   210

ABOVE The Titan probe mission as imagined in the mid to late 1980s. What eventually flew had a simpler heatshield design and an additional parachute to perform a slightly faster descent. *(ESA)*

RIGHT The CRAF mission concept. Note the penetrator flying off at the left towards the comet nucleus. A 12-sided equipment bay sits above the propellant tanks, a magnetometer boom and two RTGs are to the right, and scan platforms at the top and bottom carry instruments. *(NASA)*

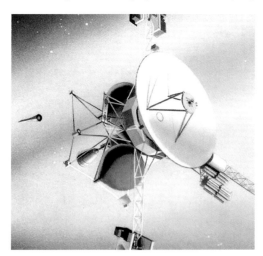

a more distant destination which demanded more propulsive capability. The original plans called for a powerful Centaur-G upper stage fuelled by cryogenic liquid hydrogen and oxygen to be carried in the Shuttle's cargo bay and, once released in Earth orbit, it would accelerate the interplanetary spacecraft onto the desired trajectory. Galileo and Ulysses, both intended for launch in May 1986, were to use this system. However, after the Challenger disaster the risks of the volatile Centaur-G propellants were deemed too great. When Galileo and Magellan both flew from Shuttles in 1989, followed by Ulysses in 1990, they all used much safer but less powerful solid-propellant rocket motors. Unfortunately these would just not be powerful enough for Cassini.

A 'Phase-A' study to develop the basic details and a more realistic cost estimate on the Titan probe began in November 1987 by the Marconi Space Systems company of Portsmouth, England. This study had to derive detailed resource demand estimates – how massive, how much power and so on – and a proof-of-concept design. Rather than the near-spherical shape of the Galileo and Pioneer Venus probes, the configuration that evolved was actually rather more squat.

A major challenge for the Europeans was that this would be their first planetary entry mission involving hypersonic aerodynamics and aerothermodynamics, disciplines only otherwise

ORGANISATIONS

Over the decades spanning Cassini-Huygens's development and operations, the organisations that made it happen have changed identities. For example, NASA's Lewis Research Center, which managed the upper stage of Cassini's launcher, became NASA Glenn in 1999. Aerospatiale, the Huygens probe contractor, became Alcatel, and more recently Thales-Alenia. Martin Marietta became Lockheed Martin; Marconi Space Systems merged with DASA (itself formerly MBB), BAe and CASA to become Astrium, which in turn is now part of Airbus... in this book I use the names pertaining at the time of mention.

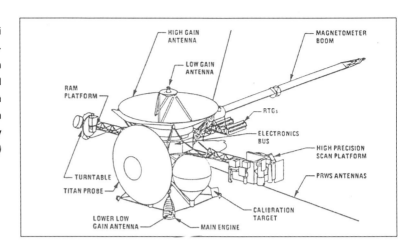

RIGHT By 1987 the 'Mariner Mark II' Cassini configuration had evolved to have teardrop-shaped tanks, and the as-yet-unnamed Titan probe was now mounted outwards. Several slender radio/plasma wave antennas stretch from the spacecraft body (later iterations placed them at the end of the magnetometer boom, but they eventually flew body-fixed, as here). *(NASA/JPL)*

encountered in ballistic missiles. The design of the heatshield in particular, and the analysis methods to estimate the heat loads it would have to endure, was within the demonstrated capability of only a few European companies. The Phase-A study suggested a high-tech lightweight carbon-carbon heatshield made by baking an epoxy-impregnated carbon fibre structure (the nose and wing edges of the Space Shuttle were made this way) together with a beryllium nose cap. Other studies by Dornier in Friedrichshafen in Germany examined materials and designs for the parachute, which would have to endure ultra-low temperatures in Titan's atmosphere and tolerate interaction with clouds of liquid methane droplets.

Meanwhile in the USA, after lobbying by various scientists, the US Congress authorised NASA to start the project. This restoration of the planetary science program was in particular stimulated by a 1987 report that underscored the value to the nation of planetary exploration, and specifically advocated a mission to the Saturn system. In 1988 NASA submitted to Congress a proposal to begin CRAF/Cassini as a formal 'new start' in 1990.

In summer 1990, work began in earnest in Europe on the Huygens probe. Two industrial consortia had bid for the contract to build it, one led by British Aerospace and the other by Aerospatiale of France.

BELOW Like Pioneer 11 and the Voyagers, Cassini would use Jupiter's gravity to slingshot its trajectory to Saturn. But the massive Cassini, which

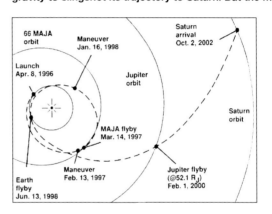

was initially envisaged as launching in 1996 to arrive in 2002, would also need a boost from an Earth flyby to get it to Jupiter. During this game of interplanetary billiards the spacecraft would fly by asteroid 80 Maja. *(NASA/JPL)*

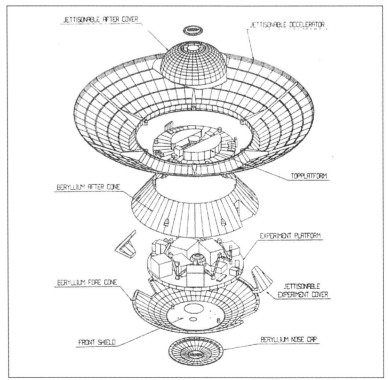

RIGHT The European Titan probe study suggested a rather 'flatter' vehicle than the probe for Galileo due to the desire to brake high in Titan's atmosphere with a large scientific payload. An exotic carbon-carbon heatshield and a lightweight beryllium structure were two of the high-tech elements considered. *(ESA)*

The selection of the experiment payload was announced in October 1990, featuring a mix of European and American instruments for the probe. The announcement for the orbiter's payload was made a few months later. There were still a few loose ends to be resolved by more detailed integration work – for example, although a science radar hadn't been selected for the probe, an engineering radar that would provide an altitude trigger might have some scientific capabilities.

The post-Voyager impression of Titan allowed that the surface might be completely covered in liquid hydrocarbons, a mix of methane and ethane (both major constituents of liquefied natural gas on Earth), and the 'model' payload for the probe included some token instrumentation in case the probe survived a landing, specifically accelerometers to record the impact or splashdown and a sensor to measure the refractive index of the liquid in order to estimate its composition.

But in 1990 new information emerged in the form of a radar echo from Titan's surface of several-hour transmissions from the giant 70m radio telescope in Goldstone, California. This was at the time the most distant radar echo ever obtained, and although interpretation of the weak reflection posed a challenge, it seemed to indicate that at least part of Titan's surface was solid. There would be no way to be sure that a simple probe would survive landing in all possible scenarios, but engineers specified that at the very least the probe should be light enough to float, and that it should have sufficient battery energy to operate for some time on the surface if it did survive.

In addition, observations of starlight being bent by Titan's atmosphere as the moon fortuitously passed in front of – 'occulted' – 28 Sagittarii in July 1989, gave credence to the presence of strong 'super-rotating' east-west (zonal) winds which had been indirectly inferred from Voyager thermal infrared measurements of atmospheric temperatures. These winds would be important in estimating how far the Huygens probe would drift during its long parachute descent, and thus might influence where a relay antenna on Cassini would have to point to receive the probe's signal. Unfortunately, neither the Voyager data nor the stellar occultation indicated whether the probe would be carried eastwards or westwards.

Another important scientific discovery around this time was the realisation that although Titan's hazy atmosphere was opaque to visible light, and thus to Voyager's cameras, near-infrared light at selected wavelengths (such as the 940nm typically used in television remote controls) could penetrate the haze. Thus not only Cassini's radar, but also its camera (fitted with a more advanced detector than that of the

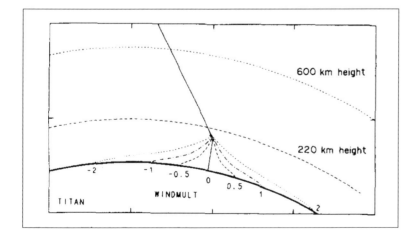

RIGHT **The Cassini configuration circa 1991. The main spacecraft had evolved to have an overall cylindrical propulsion section, with the electronics in bays just below the main antenna. It had two scan platforms, festooned with instruments and sunshades. The probe relay antenna can be seen at bottom.** *(NASA/JPL)*

Voyagers) and its near-infrared spectrometer, would be able to sense Titan's surface.

In the original Mariner Mark II concept for Cassini, the optical instrumentation would be carried on a motorised scan platform that could point independently of the spacecraft body. The instrumentation to measure energetic particles and dust would similarly be mounted on a turntable to scan them around to maximise their efficiency.

However, in 1992, the NASA planetary science budget was squeezed, in part owing to the ramp-up of spending on the International Space Station. The Mariner Mark II program, with its promise of two-for-the-price-of-one-and-a-half spacecraft, did not survive. But Cassini, with its broad and strong scientific support (and in no small measure due to the international investment in the Huygens probe) was saved. However, its cost had to be slashed. The instruments were slimmed down, and the scan platforms were deleted, the instruments instead being bolted onto the hull. Now, observations would require the whole spacecraft to be slewed around.

Another design change that was adopted at this stage was the removal of a dedicated 1.4m dish to track the Huygens probe. Instead, the large (4m) Hgh-Gain Antenna would be used to receive the signal from the probe and the large Cassini spacecraft would have to slew around to hold the line of sight. Because a bigger dish has a narrower beam, it would be necessary to receive the transmission at a longer distance so that the beam footprint would be large enough to accommodate uncertainties in the probe's position. This longer relay distance in turn meant that the probe would have to be separated from Cassini earlier, so that the separation distance had time to build up. In the end, this change would be beneficial, since the larger dish meant that the data rate could be

increased, and so the total volume of scientific data from the 2.5hr descent would be a factor of several larger than originally conceived.

An additional uncertainty in Cassini's development had been how it would reach space in the first place. The original concept assumed launch on the Space Shuttle, and in fact this option was briefly examined again in the 1992 redesign chaos. But especially after 1986, when the Centaur-G option was prohibited, Cassini was to be launched on a Titan IV, the largest expendable rocket in the US inventory. The Titan would use two large solid rocket motors for lift-off and a Centaur upper stage. The design of these solid boosters was improved for the Titan IVB to a Solid Rocket

ABOVE **The slimmed-down Cassini circa 1993, after the design was descoped significantly, with the scan platforms and probe relay antenna deleted.** *(NASA)*

Motor Upgrade (SRMU) standard that was all but essential to Cassini's success: for a while the development of these boosters was threatened by the budget axe, and in fact they were ready only just in time, with the maiden flight occurring a few months before Cassini.

Cassini's launch date vacillated during development. This was because mission studies are often overly optimistic in estimating decision and development schedules. Although the early assessments envisaged launch in 1994, the Cassini scientists were tasked to propose their instruments to the schedule laid out in the Phase-A study, with launch in April 1996 and arrival at Saturn in October 2002. Thereafter there was some shuffling between CRAF and Cassini until it was finally decided to launch Cassini in October 1997. This gave a much more comfortable schedule for the development of Cassini-Huygens and its challenging instruments. Slipping much later than October 1997, however, would impose unpalatable (and eventually intolerable) penalties as the changing planetary geometry would reduce the boost the spacecraft would get from a Jupiter flyby, with the result that the trip would take much longer.

The October 1997 opportunity was one that kept the required Deep Space Manoeuvre (DSM) to somewhat lower levels and thus saved fuel for the mission at Saturn. In fact, this plan involved no fewer than four planetary flybys to reach the target. Cassini would perversely begin its journey to the outer solar system by flying inwards towards Venus, whose gravity would deflect it out beyond Mars to perform the DSM which would aim it back at Venus for another boost. These close passes of Venus would challenge the thermal design of the spacecraft, which was optimised for the cold of Saturn, and Cassini would need to aim its High-Gain Antenna at the Sun to serve as a sunshield. The second Venus encounter, in summer 1999, would put the spacecraft on course for an Earth flyby several months later, where it would receive the kick required to arc out to Jupiter and thence to Saturn. To streamline operations costs, there was no close asteroid encounter as had been originally hoped.

The Cassini and Huygens payloads and resources

The engineering design of the Cassini Saturn orbiter and the Huygens probe as they took shape in the early and mid-1990s would be in service to their instruments, which in turn were designed to meet the scientific objectives of the mission. Their power system designs would be driven in part by the demands of the instrument payload, and their structure and configuration would be shaped to allow instruments to view targets in different directions and to sample incident particles. The ultimate purpose, of course, was to send scientific data back to Earth, and the capability of the communication system would determine that volume of data.

The actual instruments for Cassini and Huygens were selected from competitive bids submitted by teams spanning both Europe and the USA – with several times more proposals being received than there were 'slots' on the mission. The selection process was of necessity partly political, because it had to spread the selections more or less evenly across the various ESA member states as well as the USA. Of course none of the proposers knew what else would be selected, and some of the proposals covered science objectives in a slightly different way from the 'strawman' payload assumed in the Phase-A study.

For example, the Phase-A study listed a lightning detector for Huygens. While no such instrument was chosen, the selected

BELOW The interplanetary trajectory to which the Cassini and Huygens developments worked, with launch in October 1997 and arrival in 2004. The actual launch date slipped by a few days from that shown here. *(NASA/JPL)*

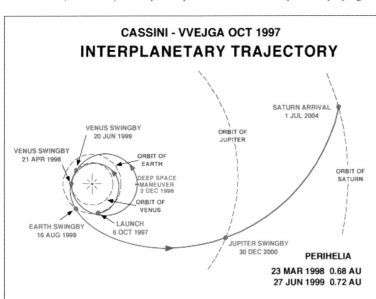

CASSINI - VVEJGA OCT 1997
INTERPLANETARY TRAJECTORY

SATURN ARRIVAL 1 JUL 2004

VENUS SWINGBY 20 JUN 1999

ORBIT OF JUPITER

VENUS SWINGBY 21 APR 1998

ORBIT OF EARTH

ORBIT OF SATURN

DEEP SPACE MANEUVER 2 DEC 1998

ORBIT OF VENUS

EARTH SWINGBY 16 AUG 1999

LAUNCH 6 OCT 1997

JUPITER SWINGBY 30 DEC 2000

PERIHELIA
23 MAR 1998 0.68 AU
27 JUN 1999 0.72 AU

Atmospheric Structure Instrument, HASI (see box) included antennas which would search for radio emissions from lightning (as would the RPWS instrument onboard the Cassini orbiter). Similarly, both HASI and SSP included accelerometers, and the different functions were examined to optimise any functional overlap (e.g. the SSP would focus on the possibly severe surface impact while HASI was tuned for the hypersonic entry deceleration). Adding the Galileo flight-spare nephelometer (cloud backscatter) instrument to the probe was also considered, but this heavy unit would require too much rework for it to be accommodated and it was decided that the DISR instrument that was selected would be able to address the aerosol optical science well enough. The scientific potential of a Huygens radar altimeter was also considered. Although a dedicated science instrument wasn't chosen, the probe would need to mark altitudes with a simple radar anyway, so it was decided to add a wire to enable the HASI instrument's signal processor to 'eavesdrop' on the radar signal and thereby extract some science data.

The scientific requirements on descent altitude and time would drive the design of the heatshield and the parachute systems, subject to ultimate constraints on the descent duration (since Cassini would fly past in just a few hours) and battery energy. Atmospheric scientists wished to analyse the light scattering and the composition of the haze and gas from as high up as possible, ideally near the 180km top of the main haze layer seen by Voyager. A rapid descent through that rarefied air would not allow much time for these first measurements, so a large parachute would be preferred. On the other hand, measurements just above and

CASSINI-HUYGENS MISSION SCIENCE OBJECTIVES

SATURN
- Determine temperature field, cloud properties and composition of the atmosphere.
- Measure global wind field, including wave and eddy components; observe synoptic cloud features and processes.
- Infer internal structure and rotation of the deep atmosphere.
- Study diurnal variations and magnetic control of ionosphere.
- Provide observational constraints (gas composition, isotope ratios, heat flux) on scenarios for the formation and evolution of Saturn.
- Investigate sources and morphology of Saturn lightning (Saturn electrostatic discharges, lightning whistlers).

TITAN
- Determine abundances of atmospheric constituents (including any noble gases); establish isotope ratios for abundant elements; constrain scenarios of formation and evolution of Titan and its atmosphere.
- Observe vertical and horizontal distributions of trace gases; search for more complex organic molecules; investigate energy sources for atmospheric chemistry; model the photochemistry of the stratosphere; study formation and composition of aerosols.
- Measure winds and global temperatures; investigate cloud physics and general circulation and seasonal effects in Titan's atmosphere; search for lightning discharges.
- Determine physical state, topography and composition of surface; infer internal structure.
- Investigate upper atmosphere, its ionization and its role as a source of neutral and ionized material for the magnetosphere of Saturn.

MAGNETOSPHERE
- Determine the configuration of the nearly axially symmetrical magnetic field and its relation to the modulation of Saturn kilometric radiation.
- Determine current systems, composition, sources and sinks of the magnetosphere's charged particles.
- Investigate wave-particle interactions and dynamics of the dayside magnetosphere and magnetotail of Saturn, and their interactions with solar wind, satellites and rings.
- Study effect of Titan's interaction with solar wind and magnetospheric plasma.
- Investigate interactions of Titan's atmosphere and exosphere with surrounding plasma.

RINGS
- Study configuration of rings and dynamic processes (gravitational, viscous, erosional and electromagnetic) responsible for ring structure.
- Map composition and size distribution of ring material.
- Investigate interrelation of rings and satellites, including embedded satellites.
- Determine dust and meteoroid distribution in ring vicinity.
- Study interactions between rings and Saturn's magnetosphere, ionosphere and atmosphere.

ICY SATELLITES
- Determine general characteristics and geological histories of satellites.
- Define mechanisms of crustal and surface modifications, both external and internal.
- Investigate compositions and distributions of surface materials, particularly dark, organic rich materials and low-melting-point condensed volatiles.
- Constrain models of satellites' bulk compositions and internal structures.
- Investigate interactions with magnetosphere and ring system and possible gas injections into the magnetosphere.

ABOVE Scientific objectives of the Cassini-Huygens mission. The set of references against which formal decisions might be made are those listed in the Announcement of Opportunity for science investigations. Note that Titan was treated separately from the other Saturnian satellites. *(NASA/JPL/ESA)*

BELOW The resource demands on the Huygens payload. *(ESA)*

Instrument	Mass (kg)	Power typical peak (Wh)	Energy (during descent) (Wh)	Typical data rate (bit/s)
HASI	6.3	15/85	38	896
GCMS	17.3	28/79	115	960
ACP	6.3	3/85	78	128
DISR	8.1	13/70	42	4800
DWE	1.9	10/18	28	10
SSP	3.9	10/11	30	704

RIGHT The Huygens
probe payload. *(ESA)*

HUYGENS PROBE INSTRUMENTS

	Measurements	Techniques	Partner Nations
Huygens Atmospheric Structure Instrument – HASI	Temp: 50–300 K; Pres: 0–2000 mbar; Grav: 1 µg–20 mg; AC E-field: 0–10 kHz, 80 dB at 2 µVm^{-1} Hz$^{-0.5}$; DC E-field: 50 dB at 40 mV/m; electrical conductivity: 10^{-15} Ω/m to ∞; relative permittivity: 1–∞; acoustic: 0–5 kHz, 90 dB at 5 mPa	Direct measurements using "laboratory" methods	Italy, Austria, Finland, Germany, France, The Netherlands, Norway, Spain, US, UK
Gas Chromatograph and Mass Spectrometer – GCMS	Mass range: 2–146 amu; Dynamic range: >10^8; Sensitivity: 10^{-12} mixing ratio; Mass resolution: 10^{-6} at 60 amu	Chromatography and mass spectrometry; 3 parallel chromatographic columns; quadrupole mass filter; 5 electron impact sources	US, Austria, France
Aerosol Collector and Pyrolyser – ACP	2 samples: 150–45 km, 30–15 km altitude	3-step pyrolysis: 20℃, 250℃, 650℃	France, Austria, Belgium, US
Descent Imager and Spectral Radiometer – DISR	Upward and downward spectra: 480–960 nm, 0.87–1.7 µm; resolution 2.4–6.3 nm; downward and side-looking images: 0.66–1 µm; solar aureole photometry: 550 nm, 939 nm; surface spectral reflectance	Spectrometry, imaging, photometry and surface illumination by lamp	US, Germany, France
Doppler Wind Experiment – DWE	(Allan Variance)1/2: 10^{-11} (in 1 s), 5×10^{-12} (in 10 s), 10^{-12} (in 100 s), corresponding to wind velocities of 2 m/s to 200 m/s, Probe spin	Doppler shift of Huygens Probe telemetry signal, signal attenuation	Germany, France, Italy, US
Surface Science Package – SSP	Gravity: 0–100 g; Tilt: ±60°; Temp: 65–100 K; thermal conductivity: 0–400 mW m^{-1}K^{-1}; speed of sound: 150–2000 m/s; liquid density: 400–700 kg m^{-3}; refractive index: 1.25–1.45	Impact acceleration; acoustic sounding, liquid relative permittivity, density and index of refraction	UK, Italy, The Netherlands, US

ideally on the unknown surface necessitated descending quickly with a smaller parachute. A compromise would be to release the large parachute after spending 10–30min at high altitude, and then descend faster on a smaller parachute in order to reach the ground nominally after 2hr 15min. Given uncertainties in the atmosphere and the aerodynamic performances (and indeed about the elevation of the surface) this meant the descent could take anywhere between 2–2.5hr. In order to optimise the measurements against this uncertainty, the

RIGHT An artist's impression of the Huygens mission sequence. During hypersonic entry, buttoned up inside the heatshield, only dynamics measurements would be made. A major design driver was the need to extract the probe from the heatshield above an altitude of about 150km in order to begin scientific measurements in Titan's main haze layer. A descent duration of 2–2.5hr was planned. *(ESA)*

probe would need to provide the experiments with some real-time altitude estimates.

Of course, the design of a scientific spacecraft is usually somewhat iterative, as the relative stiffness of the requirements are tested against the compromises that inevitably become necessary in order to satisfy budget pressures or development problems. For example, the descoping in 1992 which deleted the scan platforms also saw significant changes to the orbiter instruments and the spacecraft design (including the communications system for the probe), although the design of the probe and its payload was essentially frozen by this time.

When the scan platform was deleted, it was decided to install the suite of optical remote sensing (ORS) instruments on a pallet that formed a stiff optical bench. In addition to the boresighted instruments that spanned the ultraviolet (UVIS) through visible and near-infrared (ISS and VIMS) to the thermal infrared (CIRS), the pallet would hold the star trackers to ensure that the orientation of images relative to the stars would be precisely known as the massive spacecraft was slewed around between often long exposures or integrations at different targets.

The instruments for investigating fields and particles, which generally like to scan around continuously, were similarly bundled onto a fixed

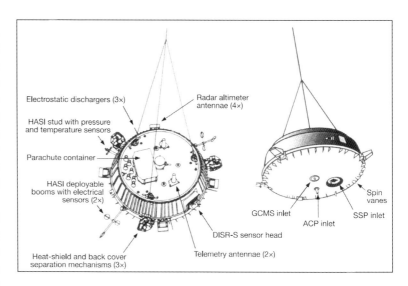

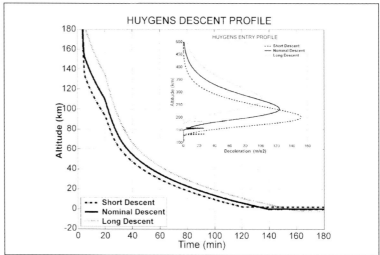

CASSINI ORBITER INSTRUMENTS

	Measurements	Techniques	Partner Nations
Optical Remote Sensing			
Composite Infrared Spectrometer – CIRS	High-resolution spectra, 7–1000 µm	Spectroscopy using 3 interferometric spectrometers	US, France, Germany, Italy, UK
Imaging Science Subsystem – ISS	Photometric images through filters, 0.2–1.1 µm	Imaging with CCD detectors; 1 wide-angle camera (61.2 mr FOV); 1 narrow-angle camera (6.1 mr FOV)	US, France, Germany, UK
Ultraviolet Imaging Spectrograph – UVIS	Spectral images, 0.055–0.190 µm; occultation photometry, 2 ms; H and D spectroscopy, 0.0002-µm resolution	Imaging spectroscopy, 2 spectrometers; hydrogen–deuterium absorption cell	US, France, Germany
Visible and Infrared Mapping Spectrometer – VIMS	Spectral images, 0.35–1.05 µm (0.073-µm resolution); 0.85–5.1 µm (0.166-µm resolution); occultation photometry	Imaging spectroscopy, 2 spectrometers	US, France, Germany, Italy
Radio Remote Sensing			
Cassini Radar – RADAR	K_u-band RADAR images (13.8 GHz); radiometry resolution less than 5 K	Synthetic aperture radar; radiometry with a microwave receiver	US, France, Italy, UK
Radio Science Instrument – RSS	K_a-, S- and X-bands; frequency, phase, timing and amplitude	X- and Ka-band transmissions to Cassini Orbiter; K_a-, S- and X-band transmissions to Earth	US, Italy
Particle Remote Sensing & In Situ Measurement			
Magnetospheric Imaging Instrument – MIMI	Image energetic neutrals and ions at less than 10 keV to 8 MeV per nucleon; composition, 10–265 keV/e; charge state; directional flux; mass spec: 20 keV to 130 MeV ions; 15 keV to greater than 11 MeV electrons, directional flux	Particle detection and imaging; ion-neutral camera (time-of-flight, total energy detector); charge energy mass spectrometer; solid-state detectors with magnetic focusing telescope and aperture-controlled ~45° FOV	US, France, Germany
In Situ Measurement			
Cassini Plasma Spectrometer – CAPS	Particle energy/charge, 0.7–30,000 eV/e; 1–50,000 eV/e	Particle detection and spectroscopy; electron spectrometer; ion-mass spectrometer; ion-beam spectrometer	US, Finland, France, Hungary, Norway, UK
Cosmic Dust Analyzer – CDA	Directional flux and mass of dust particles in the range 10^{-16}–10^{-6} g	Impact-induced currents	Germany, Czech Republic, France, The Netherlands, Norway, UK, US
Dual Technique Magnetometer – MAG	B DC to 4 Hz up to 256 nT; scalar field DC to 20 Hz up to 44,000 nT	Magnetic field measurement; flux gate magnetometer; vector–scalar magnetometer	UK, Germany, US
Ion and Neutral Mass Spectrometer – INMS	Fluxes of +ions and neutrals in mass range 1–66 amu	Mass spectrometry	US, Germany
Radio and Plasma Wave Science – RPWS	E, 10 Hz–2MHz; B, 1 Hz–20 kHz; plasma density	Radio frequency receivers; 3 electric dipole antennas; 3 magnetic search coils; Langmuir probe current	US, Austria, France, The Netherlands, Sweden, UK

structure instead of the turntable envisaged by the Phase-A study. As a compromise, rotating elements were incorporated into instruments for which scanning was particularly important, like CDA and MIMI-LEMMS.

The figures listed in this section represent the as-built payload numbers, but are fairly representative of the values against which the designs were made. Normally, an allocation for the power and mass of an instrument is made early in the process and some margin is retained at the project level to be applied as development problems occur. But this is a bit of a game because the instrument builders know there is a margin and that their jobs will be easier if they get some of it. Unusually, the Cassini orbiter's payload development actually tried an innovative approach wherein there was no centrally held margin, but the instruments had the flexibility to 'trade' resources – e.g. 1kg of mass allocation in return for a couple of watts of power, or 10kbps of data rate in return for $200,000 of fiscal year 1993 funds. Under the trading system (in which 29 trades were made, mostly for dollars and mass) the total instrument cost growth was less than 1% and the total payload mass was 7% less than its allocation. This result is in stark contrast to the typical 50–100% increases in these resources on past

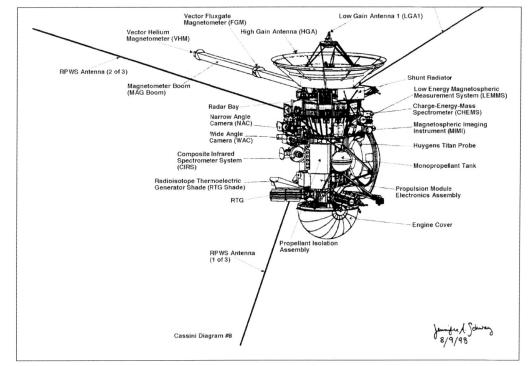

ABOVE The Cassini orbiter payload. *(NASA/JPL)*

RIGHT The overall configuration of Cassini, with many of the payload elements identified. This view shows the main engine cover in its deployed position. *(NASA/JPL)*

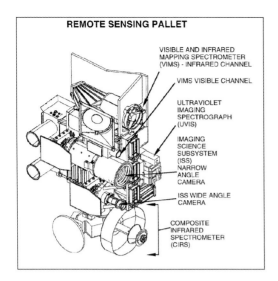

REMOTE SENSING PALLET

VISIBLE AND INFRARED MAPPING SPECTROMETER (VIMS) - INFRARED CHANNEL

VIMS VISIBLE CHANNEL

ULTRAVIOLET IMAGING SPECTROGRAPH (UVIS)

IMAGING SCIENCE SUBSYSTEM (ISS) NARROW ANGLE CAMERA

ISS WIDE ANGLE CAMERA

COMPOSITE INFRARED SPECTROMETER (CIRS)

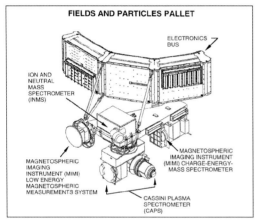

FIELDS AND PARTICLES PALLET

ELECTRONICS BUS

ION AND NEUTRAL MASS SPECTROMETER (INMS)

MAGNETOSPHERIC IMAGING INSTRUMENT (MIMI) CHARGE-ENERGY-MASS SPECTROMETER

MAGNETOSPHERIC IMAGING INSTRUMENT (MIMI) LOW ENERGY MAGNETOSPHERIC MEASUREMENTS SYSTEM

CASSINI PLASMA SPECTROMETER (CAPS)

LEFT Following the deletion of the scan platform in 1992, the optical instruments were mounted on a fixed pallet that served as an optical bench, bolted onto the side of the spacecraft. In addition to the science instruments, this carried the two stellar reference units (SRUs or 'star trackers') which can be seen at the left. *(NASA/JPL)*

BELOW LEFT Several of the fields and particles instruments were installed on a fixed structure that was attached to some of the electronics boxes on the upper part of Cassini. The RPWS with its long antennas was mounted separately, as was the CDA, which had a turntable of its own in order to scan in different directions. *(NASA/JPL)*

BELOW The mass and power breakdown of the Cassini orbiter payload and subsystems. *(NASA/JPL)*

SPACECRAFT SUBSYSTEMS MASS AND POWER

Subsystem or Component	Mass, kilograms	Power, watts	Comments on Power Usage
Engineering Subsystems			
Structure Subsystem	272.6	0.0	
Radio Frequency Subsystem	45.7	80.1	During downlinking of data
Power and Pyrotechnics Subsystem	216.0	39.1	Shortly after launch
Command and Data Subsystem	29.1	52.6	With both strings operating
Attitude and Articulation Control Subsystem	150.5	115.3	During spindown of spacecraft
(Engine Gimbal Actuator)		31.0	During main engine burns
Cabling Subsystem	135.1	15.1	Maximum calculated power loss
Propulsion Module Subsystem (dry)*	495.9	97.7	During main engine burns
Temperature Control Subsystem	76.6	117.8	During bipropellant warmup
		6.0	Temperature fluctuation allowance
		2.0	Radiation and aging allowance
		20.0	Operating margin allowance
Mechanical Devices Subsystem	87.7	0.0	
Packaging Subsystem	73.2	0.0	
Solid-State Recorders	31.5	16.4	Both recorders reading and writing
Antenna Subsystem	113.9	0.0	
Orbiter Radioisotope Heater Subsystem	3.8	0.0	
Science Instrument Purge Subsystem	3.1	0.0	
System Assembly Hardware Subsystem	21.7	0.0	
Total Engineering Mass*	1756.6		
Science Instruments			
Radio Frequency Instrument Subsystem	14.4	82.3	Both S-band and K$_a$-band operating
Dual Technique Magnetometer	8.8	12.4	Both scalar and vector operations
Science Calibration Subsystem	2.2	44.0	In magnetometer calibration mode
Imaging Science (narrow-angle camera)	30.6	28.6	Active and operating
Imaging Science (wide-angle camera)	25.9	30.7	Active and operating
Visible and Infrared Mapping Spectrometer	37.1	24.6	In imaging mode
Radio and Plasma Wave Science	37.7	17.5	During wideband operations
Ion and Neutral Mass Spectrometer	10.3	26.6	Neutral Mass Spectrometer operating
Magnetospheric Imaging Instrument	29.0	23.4	High-power operations
Cosmic Dust Analyzer	16.8	19.3	Operating with articulation
Cassini Radar	43.3	108.4	Operating in imaging mode
Cassini Plasma Spectrometer	23.8	19.2	Operating with articulation
Ultraviolet Imaging Spectrograph	15.5	14.6	On in sleep state
Composite Infrared Spectrometer	43.0	43.3	Active and operating
Total Science Mass	338.2		
Huygens Probe (including Probe support)	350.0	249.8	During Probe checkouts
Launch Vehicle Adaptor Mass	136.0	0.0	
TOTALS*	2580.7	680.5	Power available at middle of tour

* Note that the masses given above do not include approximately 3141 kilograms of propellant and pressurant.

missions. Thanks to this market-based system, the entire payload flew.

Although NASA paid for all of the US instruments on Cassini and Huygens, the European instruments (and European team members or other contributions to US-led instruments) were funded not by ESA but by the individual European governments. While NASA continued to have political challenges of its own, funding shortfalls which NASA and ESA could not control led to delays in instrument development. But at least ESA's relatively cumbersome decision-making process meant that once the mission was approved in 1990, the development of the probe itself had relatively secure funding.

The systems and payloads of the probe and the orbiter raced to meet their technological and budgetary challenges in order to meet the launch scheduled for October 1997, but because the probe needed to be integrated first, it and its instruments had the tightest development schedule. So we shall consider these first.

Chapter Two

The Huygens probe

────●────

The Huygens probe presented new challenges in planetary exploration and was Europe's first foray to another world. In-situ investigation of a dense, cold atmosphere after seven years in space would demand careful design and testing.

OPPOSITE Integration of the Huygens probe in Ottobrun, Germany. Assembly and test of a multinational spacecraft takes meticulous planning, with attention to infrastructure details. The fixtures (from Austria) used to hold the back cover and (French) front shield had to be designed years in advance, noting the bearing capacity of the facility floor, the size of the doors, and so on. *(ESA)*

Huygens's payload

Six instruments were selected which spanned the Titan science objectives outlined for Cassini and the Huygens probe, namely:

- Determination of the atmospheric composition.
- Investigation of energy sources driving the atmospheric chemistry.
- Investigation of aerosol properties and cloud physics.
- Measurement of global winds and temperature.
- Determination of surface properties and internal structure.
- Investigation of the upper atmosphere and ionosphere.

In addition to the six experiments (DWE, DISR, ACP, GCMS, HASI and SSP), the project included three Interdisciplinary Scientists who were selected to help to coordinate investigations across the instruments, and between the probe instruments and the orbiter investigations of Titan.

The ***Aerosol Collector and Pyrolyser*** was to trap haze particles suspended in Titan's atmosphere using a deployable sampling device – essentially a thimble of 0.4mm stainless steel mesh that was extended out from the bottom of the probe, through which air was drawn in by a 50,000rpm fan (much like a vacuum cleaner). After accumulating a sample for some tens of minutes as the probe was descending from 80–32km, the filter was pulled up inside the instrument and a gate valve slid across to seal the sample in an oven. Nitrogen gas (isotopically labelled N-15) from a small (32mm diameter) stainless steel tank was used to flush vapours from the sample through a narrow pipe to the GCMS instrument for analysis, then the sample was heated to 250°C and these less-volatile vapours were flushed to GCMS, and then finally the sample was cooked to 600°C to break down (pyrolyse) the bigger compounds for GCMS to analyse.

The cycle was repeated with a second deployment of the filter in the altitude range 22–17km. (Initially two separate filters were planned but this mechanically sophisticated instrument, drawing on experience of a similar instrument flown on a Russian Venus mission a few years earlier, had to be descoped.)

The instrument was led by the Service d'Aeronomie of the Centre National de la Recherche Scientifique in Verrieres-le-Buisson, France. Much of the electronics was provided by Austria.

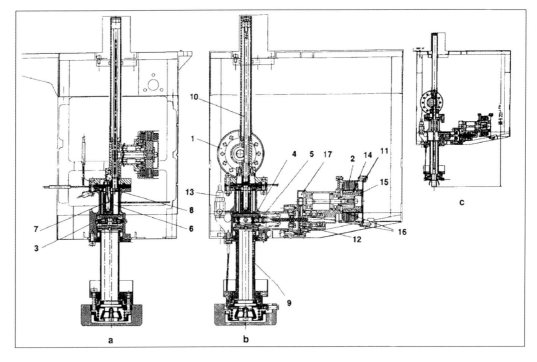

RIGHT The ACP internal mechanism was complex. A rack mechanism advanced a filter down to the inlet and after sucking gas through the filter with a pump, retracted it into an oven for heating.
(Service d'Aeronomie)

Since the ACP was basically a chemical processor without analysis instrumentation, it was quite simple from a data-handling point of view. An 8-bit 80C85 microcontroller which ran at 4MHz and executed a program of 12 kBytes was to sequence the motors and heaters and read the sensors. Communications with the probe were handled by two ACTEL-1020A field programmable gate arrays (FPGA).

The box structure was aluminium, but many internal mechanisms and housings were made of titanium for stiffness. The filter was retracted through a stroke of 120mm by a stepper motor driving a rack and pinion mechanism. A magnet locked the filter in place. The gate valve used a brushless DC motor to drive a nut and screw mechanism which produced a large force (500–900N) to seal the oven against a fluorosilicone O-ring.

The 850g pump developed by Technofan had a pair of rotors and used a brushless three-phase synchronous motor by Norcroft that drew up to 69W. As the unit only needed to operate for ten hours (including tests) an unlubricated bearing was planned, but jamming during development tests prompted the project to use a Duroid (glass-reinforced Teflon) cage around stainless steel ball bearings with molybdenum disulphide solid lubricant.

The instrument was sealed against contamination (both the inlet and an exhaust tube downstream of the fan) by spring-loaded caps held in place by silver-tin solder. After parachute deployment and heatshield release, the caps were electrically heated (53W) to melt the solder within a few minutes and hence open the tubes.

The **Gas Chromatograph/Mass Spectrometer** addressed the core question of what gases were in Titan's atmosphere. Gas was forced by the dynamic pressure of descent into an inlet pipe that exhausted to the top platform. A tiny amount was drawn in from this pipe, through valves and flow restrictors to several destinations. Some gas from early in descent was piped to an enrichment cell, where metal 'getter' material absorbed the nitrogen that forms the bulk of the atmosphere, to facilitate the detection of traces of noble gases such as argon which would provide clues to the origins of the moon's atmosphere.

LEFT The ACP experiment. The inlet cover at the top was released by electrical heating to melt a solder seal, to allow a spring to push off the cover. The electronics cards were mounted at the side of the box. *(Service d'Aeronomie)*

LEFT The ACP pump unit was a fan, spun at 60,000rpm to draw air in through a filter. The unit had to operate after spending seven years in space. *(Service d'Aeronomie)*

LEFT The GCMS spare at the Goddard Space Flight Center, with its inlet assembly at the top. Because the instrument had high voltages, the interior was maintained at elevated pressure to suppress arcing, making the housing resemble a scuba tank. The length of the unit required that it be mounted through the main experiment platform structure on the probe, instead of being bolted to one side or the other. *(Author)*

RIGHT The plumbing diagram of the GCMS. Routing of sample gas was achieved with more than 20 microvalves, passing gas from the atmospheric inlet (or the ACP instrument) into the mass spectrometer either directly or via chromatography columns or a noble gas enrichment cell. Electrically charged molecules produced from the gas by an ion source were separated in a quadrupole mass analyser. *(NASA/GSFC)*

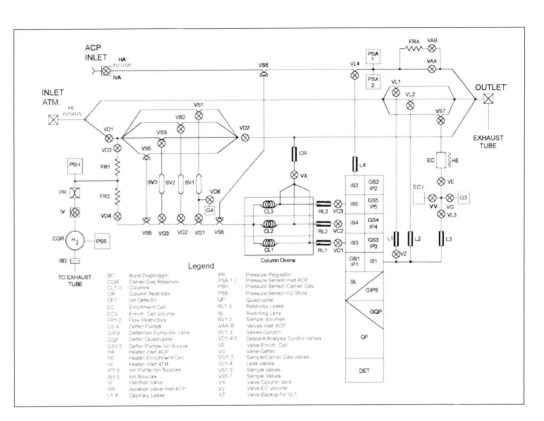

BELOW LEFT The internals of GCMS. The cylindrical mass analyser containing the quadrupole rods (in fact, the same rods as the Cassini INMS instrument) is shown at the left, with the gas handling hardware on the platform above. To the right are the electronics installed around the mass analyser. *(NASA/GSFC)*

BELOW CENTRE A close-up view of the GCMS gas handling hardware. The inlet tube is at the top, plugged with a blob of white sealant. Square boxes hold a gettering compound to absorb gases and maintain the internal vacuum. A dozen gas valves are arranged to the right. The hydrogen gas tank and pressure regulator for the gas chromatograph are at the lower left. The complexity of the plumbing is evident. *(Author)*

BELOW RIGHT A microvalve of the GCMS. The critical development of these valves by a small company was a constant worry for the GCMS team. One Goddard Space Flight Center engineer recalls having to take an overnight flight back from California on Christmas Eve with a handful of the precious newly made valves in order to maintain the GCMS schedule. *(NASA/GSFC)*

Some samples, including material from the ACP, were blown through one of three Gas Chromatograph columns which separated compounds by their affinity for the coatings on the walls of the columns. The combination of GC and MS was a powerful means of disentangling what was expected to be a complex mix of materials – overall the mass spectrometer was similar to that supplied by the same group for the Galileo probe, but the addition of the Gas Chromatograph made the whole system rather more capable and complex.

Additionally, for much of the descent and time on the surface, gas was introduced directly into the MS which used heated filament ion sources to ionise it. Quadrupole electrodes driven by high voltage radio frequency signals then separated the ions having various mass-to-charge ratios and identified atoms and molecules up to 141 Daltons – the Dalton being the unified atomic mass unit which is approximately the mass of a single proton in an atomic nucleus. The quadrupole electrodes were identical to those on the orbiter's Ion and Neutral Mass Spectrometer. The GCMS development was led by NASA's Goddard Space Flight Center in Greenbelt, Maryland, USA.

The vacuum inside the Mass Spectrometer was maintained by getters (a sintered titanium-molybdenum powder that absorbs hydrogen and nitrogen gas) and ion pumps. Ironically, owing to the high voltages used to drive the quadrupole, the electronics of the GCMS had to be kept under pressure (0.12 MPa dry nitrogen) to avoid electrical discharge arcing, therefore the housing of the instrument was overall rather formidable. The instrument was sealed during the cruise by ceramic break-off caps that were released by pyrotechnic actuators fired by the probe electronics.

The elaborate plumbing of this instrument needed two dozen valves to control the different functions. The inlet pipe was heated to prevent any droplets of material from blocking the inlet tube (as happened on the Pioneer Venus probe in 1979). A 2m long 0.7mm diameter GC tube was employed to separate carbon monoxide and nitrogen (which have the same atomic mass); the other two GC columns were to separate large and small nitrile and hydrocarbon molecules respectively, and were 10m and

14m long with only 0.18mm internal diameters. Hydrogen carrier gas was used, owing to its efficient storage and pumping, with the 3 standard litres needed being stored in a 100g hydride metal alloy in a stainless steel reservoir sealed for the cruise by a diaphragm.

The **Descent Imager/Spectral Radiometer** had 13 optical apertures aimed in different directions, and in the wavelength range 350–1,700nm in order to measure the intensity and spectrum of light as the probe descended and to image any cloud structure and surface features. Using the probe's rotation, the imagers could build up a mosaic of pictures around the landing site. A side-looking visible imager was to observe the horizon and the underside of any cloud deck.

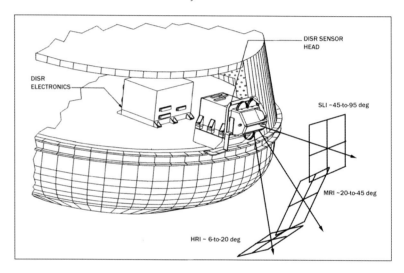

ABOVE **The DISR instrument looked out of the side of the probe. The three imager fields of view (high resolution, medium resolution, and side-looking) are shown. The Sun sensor and several photometers and spectrometers looked in upwards directions.** (U. of Arizona)

BELOW **Fibre optic bundles guided light from the different optical apertures onto the CCD detector of the DISR instrument.** (U. of Arizona)

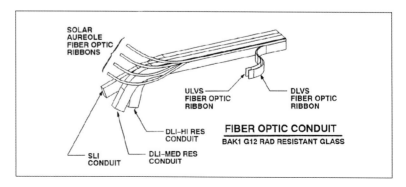

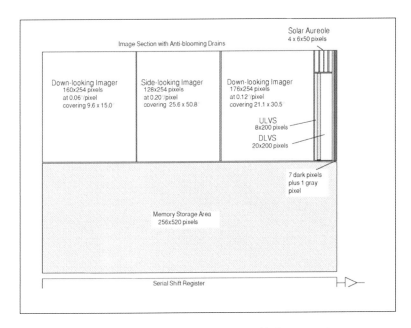

In many respects this instrument was inspired by simple radiometer instruments on Pioneer Venus, with the intent to understand how Titan's haze absorbed and scattered sunlight and thus controlled its climate via greenhouse and anti-greenhouse effects, with imaging added as a bonus (remember, when Huygens was designed, the data rate from the

probe to the orbiter was rather lower than it ended up being and so the number of images would have been severely limited).

The instrument used a rather novel architecture, mapping different fields of view (with different optical systems) to different portions of a single Charge-Couple Device (CCD) image detector. Although modern cameras have CCDs holding many megapixels, in 1990 when DISR was proposed, a quarter megapixel (520x512 pixels, each 23μm) chip was the state of the art, and only a couple of CCD imagers had been built for deep space missions (Giotto and Galileo). The mapping was performed by a system of optical fibres made from special radiation-hard glass, with the different fibre bundles being fed from lens/filter heads oriented in various directions.

The CCD subsystem was developed at the Max-Planck Institute Lindau in Germany (which had led the Giotto camera) using a CCD chip by LORAL Fairchild. There was no mechanical shutter, the images were rapidly shifted electronically ('frame transfer') to a blanked storage area of the 12x12mm chip for readout and digitisation – a process which took 2.2sec.

ABOVE The DISR instrument used a single CCD detector for several different functions, with different areas of the chip mapped by the fibre bundles to their optical apertures. In 1990, when the instrument was proposed, the quarter-megapixel CCD chip was near the state of the art in digital cameras. *(U. of Arizona)*

RIGHT This block diagram highlights the complexity of combining so many functions in a single instrument. The shading hints at the organisational challenge common to most Cassini-Huygens instruments, with the various hardware elements being supplied by different countries.
(U. of Arizona)

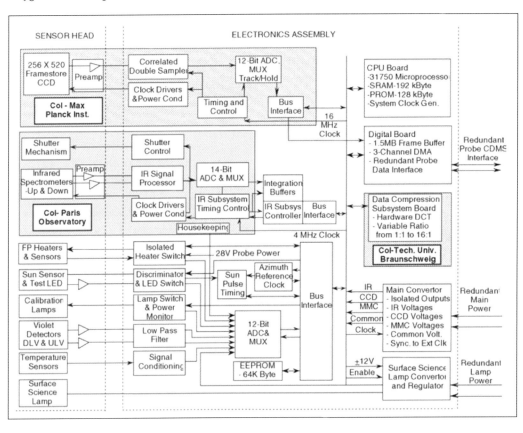

Because the data rate would be limited, the instrument had to compress its information. A dedicated system was to do the Discrete Cosine Transformation (DCT) and Huffman coding (essentially similar to JPEG compression) after the 12-bit raw pixel readings were compressed to 8 bits using a square-root encoding look-up table. The compressor used an 80C86 microprocessor, an SCT3200 DCT chip, and eight ACTEL FPGAs typically performing compressions in the range 3:1 to 8:1.

Although Titan receives 100 times less light than Earth due to its greater distance from the Sun and there is another factor of 10 reduction near the surface owing to absorption by the haze, the illumination is still several hundred times brighter than full moonlight on Earth. But the low light levels and the rotation of the probe (which limited the exposure time) required the optical system to be fairly 'fast' with a focal ratio (focal distance to lens diameter) of f/2.5. The angular resolution of the high-resolution imager was 0.06°, which is comparable to the naked human eye.

Because there was predicted to be very little violet, blue, or even green light near Titan's surface due to absorption by the haze, the imagers operated over a wide bandpass between about 600nm (orange) and 1,000nm (near-infrared). Although the images themselves were monochrome (or rather, panchromatic), colour information could be added from measurements with the upward-looking and downward-looking visible spectrometers that used diffraction gratings to measure 480–960nm in their first order dispersion and used 200-pixel long blocks of the CCD. Violet light levels, particularly sensitive to the amount of (red) Titan haze, were measured by upward-looking and downward-looking photometers that employed silicon photodiode detectors. Infrared spectra were measured in the range

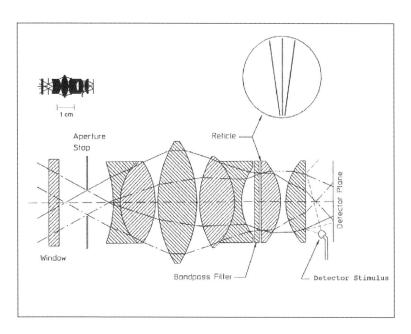

ABOVE Each of the DISR sensors had an optical system to focus, collimate, and filter the light onto a detector. This shows the Sun sensor, which used a lens arrangement narrow filter at 940nm to isolate the wavelength of light least affected by the atmosphere, and a set of three slits to cause the Sun to produce three pulses on a photodiode detector as the probe rotated, with the pulse timing indicating the Sun's position. There was an internal LED to stimulate the detector for aliveness and sensitivity tests during cruise checkouts. *(U. of Arizona)*

LEFT The sensor head of the DISR instrument, showing a black insulating anti-reflective coating on the exposed part at the right. The copper strap in the centre was to conduct heat away to cool the detector. The large Canon D-connectors at top left have connector savers in place for testing, so that the gold coating on the connector pins was not worn out by repeated mating and demating prior to flight. *(C. See/U. of Arizona)*

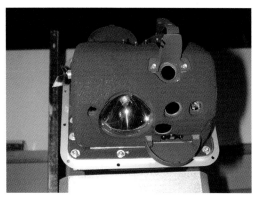

RIGHT A front lower view of the sensor head of the DISR instrument. The gold reflector of the surface science lamp is at the left, while the three circular imager apertures are seen at the right (the SLI aperture has a sunshade over it). Semicircular ('bear's ear') baffles collimated the light to the spectrometers. *(C. See/U. of Arizona)*

800–1,700nm with a separate linear array of 150 indium gallium arsenide photodiodes and a vibrating shutter.

Additional fibre channels, also feeding the CCD but with polarising filters, measured the light intensity near the line to the Sun (named the solar 'aureole') to measure the way in which the haze scattered light and hence estimate the haze particle size.

DISR had a Sun sensor which measured the solar position by the timing of pulses of light seen through three slits as the probe rotated. With the azimuth information from this sensor, DISR could optimise the timing of the aureole and imaging measurements to obtain the desired angles.

The detectors and optical apertures were mounted in a sensor head box that poked out of the side of the probe. It was equipped with

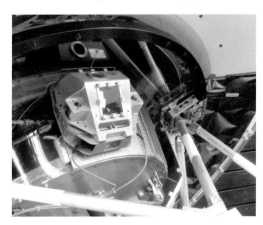

thermal straps to cool the detector to minimise 'noise' in the electronics, as well as a 1kg tungsten radiation shield to protect the sensitive detector from space radiation and especially from neutrons emitted by the RTGs and heaters on Cassini. The sensor head also incorporated calibration lamps to check the system response during cruise, as well as a 20W surface science lamp to illuminate a patch of ground with 'white light' (i.e. unfiltered by the hazy atmosphere above) in order to measure the reflectance spectrum with a good signal-to-noise just prior to impact. The sensor head was connected to a large electronics box to perform the data handling and sequencing of measurements.

The relatively elaborate data processing was handled by an MA31750 processor which ran at 12MHz and was capable of 1.6 million instructions/sec (75% of which were required by DISR). 128kB of program random access memory (RAM) was copied from a fixed Programmable Read-Only-Memory (PROM), with updates in a 64kB Electrically Erasable PROM (EEPROM – at the time of Huygens development, space-qualified EEPROMs and 12-bit analogue-to-digital converters were just becoming available). The software was coded in Ada.

The instrument was led by the University of Arizona in Tucson, with design and fabrication by the Martin Marietta company in Denver, Colorado. The image detector and data compressor was sourced in Germany and an infrared spectrometer was developed by the Paris Observatory.

The **Huygens Atmospheric Structure Instrument** investigated the physical properties of the atmosphere. During the hypersonic entry, an accelerometer inside the heatshield recorded the deceleration to permit the density of the upper atmosphere to be calculated. Then during parachute descent, the pressure and temperature were measured directly. A set of electrodes on booms, the Permittivity and Wave Analyser (PWA), measured the ion and electron conductivity of the atmosphere and listened for electromagnetic wave activity such as emissions from lightning. The instrument also processed the signal from the radar altimeter in order to obtain information about the topography, roughness and electrical properties of the surface.

RIGHT Not so much an instrument as an infestation, HASI featured several units attached to different parts of the probe. The accelerometers were placed at the centre to minimise rotation effects on the signal; the pressure port and temperature sensors on a small stub boom to get them out of the boundary layer of stagnant airflow near the skin; and the two PWA booms that swung out to measure electric fields. *(HASI Team)*

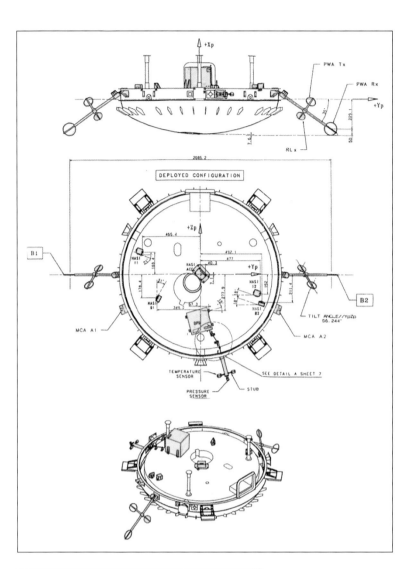

On reaching Titan's surface, the instrument was to measure the impact and the electrical permittivity of the material, plus, in the case of a splashdown, any wave motion.

The instrument development was led by the Paris Observatory (initially the principal investigator was at the University of Rome but, like Cassini himself, he moved to Paris) with hardware contributions from Finland, the UK, Austria, Spain, and ESA.

The accelerometers were mounted as near as possible to the centre of mass of the probe-and-heatshield configuration, so that rotational contributions to the acceleration signal would be minimised. The most sensitive sensor, mounted on the axis of symmetry of the probe, was a Sundstrand QA-2000-030 servo accelerometer which could sense a few micro-g of acceleration. Small piezoresistive sensors (Endevco 7264A-2000T) measured accelerations in three axes at a much lower sensitivity.

In order to maximise the accuracy of the pressure measurement as the probe descended through air that got progressively denser by

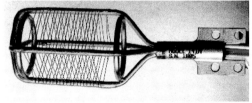

ABOVE RIGHT The atmospheric temperature was sensed using a long, thin platinum resistance wire to minimise its response time. *(M. Leese/HASI Team)*

RIGHT The pipe at the lower right conveyed the pressure from the Kiel tube to a spider-like manifold. Because the pressure changed by a factor of several hundred during descent, the pressure was measured by several sensors which had overlapping ranges and were attached to a thermally stable mounting plate. *(Author)*

a factor of 1,000, sets of pressure sensors
with three ranges were employed: 0–400kPa,
0–1,200kPa and 0–1,600kPa.

The temperature sensor used a fine 99.999%
pure platinum wire as a resistance thermometer.
This 2m long, 0.1mm diameter wire was
optimised to give the sensor a short response
time in the tenuous upper atmosphere. It was
mounted via a thin glass insulation layer on a
platinum-rhodium frame.

The **Doppler Wind Experiment** used a pair
of ultra stable oscillators, one (transmit, TUSO)
on the probe and one on the orbiter (receive,
RUSO) to provide the radio relay link a stable
transmit carrier frequency and a precise
reference against which to measure its Doppler
shift. This shift would indicate the relative motion
of the probe and orbiter along the line of sight.
Given tracking of the orbiter's motion relative to
Titan, the motion of Huygens relative to Titan's
surface would provide a profile of the winds
during the descent.

The investigation was led by the University of
Bonn in Germany. Each USO – built by Daimler-
Benz Aerospace (DASA) around a rubidium cell
supplied by Efratom Elektronik – was installed in
a 1.9kg 17x12x12cm box that was coated with
Chemglaze Z306 thermal paint.

The rubidium 'physics package' was
essentially an atomic clock using a vapour
discharge lamp and a resonance cell. A radio
frequency voltage applied to the cell from
a quartz oscillator would resonate, via tiny
variation of light transmission through the cell,
with the electronic transitions of the rubidium
atoms possessing a precise and stable absolute
frequency. Once locked to these transitions,
the USO (it was the first time this type was
used on a deep space probe) could maintain a
frequency accuracy within 2 parts in 10 billion
(2×10^{-10}) of its nominal value with a stability of
several parts in a trillion (10^{-12}) over periods

ABOVE The core of the ultra stable oscillator was a low-pressure gas discharge cell that drove a quartz oscillator. It was installed within a thermally insulating housing to provide maximum stability. (M. Bird/DASA)

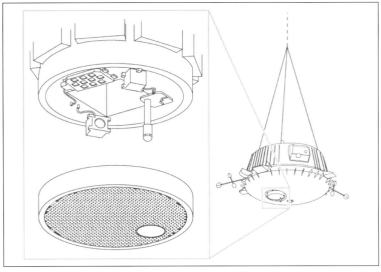

ABOVE The sensors of the Surface Science Package were mounted on a cylindrical 'Top Hat' structure on the base of the probe so that the sensors would be immersed in the event of a liquid landing. The penetrometer projected ~5cm ahead of the base. Also shown at the right is the metal mesh cover (with hole for the penetrometer) to suppress any electrical interference or damage from possible lightning. (James Garry)

of minutes. The 10MHz frequency output of the USO was upconverted to the 2,040MHz S-band probe relay signal on one of the probe's two radio links. Comparing this signal with the RUSO onboard Cassini would indicate the frequency difference.

The **Surface Science Package** carried a number of sensors to determine the physical properties and composition of Titan's surface, particularly in the case of a landing in liquid. The original proposal included an X-ray fluorescence spectrometer (similar to instruments used on the Moon and Mars) to identify elements in rocks, but this had to be deleted due to budget limitations.

Simple physical property measurements (refractive index and thermal conductivity) could be made in seconds and would be diagnostic of whether Titan's seas were composed predominantly of methane or ethane. A small echo-sounder or sonar – a first for space exploration! – would operate at 15kHz (just beyond most people's hearing) to measure the depth of the sea, supplemented by a second set of sensors to measure the speed of sound. A permittivity sensor was to measure

the dielectric constant of the liquid, this being another diagnostic of composition and useful to compare with data from Cassini's radar. An accelerometer would measure the deceleration during impact or splashdown. If the surface was solid, a penetrometer projecting from the base of the probe would make a local, higher resolution measurement of the impact resistance. And tilt sensors would measure the probe orientation and any wave motion. Some of the sensors would make limited measurements during the descent (e.g. the

LEFT The engineering model of the SSP 'Top Hat', seen from the bottom. Silvered film packs of foam insulation are visible at top. Inside the square cavity can be seen a parallel-plate capacitor to measure the sea dielectric constant, cylindrical thermal sensors, the curved prism of the refractometer, and a white density sensor. Around the perimeter of the front plate (clockwise from the top) are the two speed-of-sound transducers, acoustic sounder, and penetrometer. (Author)

RIGHT The glass fibre reinforced plastic structure of the 'Top Hat'. In its Titan orientation the box cavity was presented downwards to the surface and the cut-out dome provided for secure attachment to the Huygens experiment platform. The vent hole of the cavity was mated to a vent tube to allow the cavity to quickly fill on immersion. The structure was designed to be stiff yet lightweight, while having low thermal conductivity. A crack that developed in vibration testing led to hurried repairs during development.
(M. Leese/SSP Team)

speed of sound in air and variation in tilt as the craft swung under its parachute), as well as after impact.

The instrument was led by the University of Kent in Canterbury, England (which had built Giotto's dust impact detector), with the support of the Rutherford Appleton Laboratory near Oxford. The acoustic sensors were supplied by ESA and the thermal conductivity sensor by the Space Research Institute in Warsaw, Poland.

Most of the sensors required access to Titan's surface, and if it was liquid, to be

immersed in it. SSP therefore formed a sensor cavity (with a vent pipe, so that it would fill with liquid) on the base of the probe. This used a lightweight glass fibre plastic structure supported with thermal insulation foam to mount the sensors to measure the density, speed of sound, refractive index, thermal conductivity and electrical permittivity of the liquid. Also mounted on this structure, nicknamed the 'Top Hat', was the downward-looking acoustic sounder and the penetrometer to measure the hardness of the ground in a solid impact.

These custom sensors all had to function at ambient conditions of 94K or minus 179°C. The thermal conductivity and permittivity sensors were heated platinum wires, and a parallel plate capacitor. The density sensor used a small foam float on a tiny strain gauge cantilever, and required good temperature compensation for data analysis. The acoustic sensors and the penetrometer used piezoelectric ceramics (in the case of the penetrometer, a lead zirconate titanate PZT-5A) which could function happily at low temperatures. The ingenious refractometer used a curved sapphire prism to measure the liquid index by means of total internal reflection – the angle being measured using a photodiode array recessed into the insulation so that it was not quite as cold as the liquid itself.

The sensor suite included some off-the-shelf items in the internal electronics box – such as an Endevco 2271AM20 piezoelectric accelerometer to record Huygens' impact with the surface; and a pair of LU-211 electrolytic tilt sensors that used a slug of potassium chloride solution in a glass vial, their position being sensed using shaped electrodes to indicate vertical to within a fraction of a degree for measuring swing under the parachute and motion on possible waves.

SSP used a MAS-281 radiation-hardened silicon-on-sapphire 1750A processor, coded in assembly language to interrogate the sensors and format their output for the probe CDMS. One important job was to quickly summarise and transmit key readings into an 'impact packet' immediately upon landing, just in case the probe rapidly succumbed to waves or some other unknown hazard of the surface of Titan.

BELOW The penetrometer, built by the author, was the first part of the Huygens probe to make contact with the surface. The hemispherical titanium bolt at the right was 14mm in diameter and squeezed a white disk of PZT-5A piezoelectric ceramic to generate a force history signal upon impact. A spare unit is on display in the Science Museum in London.
(M. Leese/SSP Team)

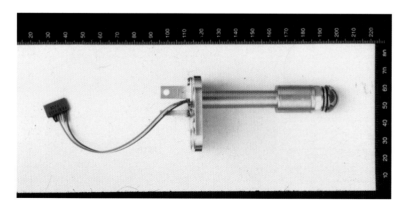

Huygens probe design and construction

The job of the Huygens probe was to convey its instruments to Titan and support them with electrical power and a benign thermal environment during their operation and data transmission to Cassini. The probe design would have to take into account the environment on the ground (on Earth), conditions in interplanetary space from the vicinity of Venus all the way out to Saturn, and the hypersonic entry into and descent through the frigid, dense atmosphere of Titan. Its structure had to tolerate the brutal g-load, vibration, and acoustic pummelling of launch on the Titan IV rocket, as well as the g-load during Titan entry. A series of pyrotechnic charges had to break the probe out of its entry cocoon with split-second timing and deploy the first of three parachutes while travelling at supersonic speed. And the electronic systems had to operate reliably without intervention from Earth, and also support occasional checkouts and updates during seven years in space.

All of these aspects had to be thought through in the Huygens system proposals provided to ESA. After Aerospatiale was chosen as the prime contractor, the industrial Phase-B activities began with a kick-off meeting in January 1991 at Aerospatiale's premises in Cannes, France – at the time that public attention was focused on the start of the First Gulf War!

ESA itself does almost no hardware construction. Instead, it specifies and manages the development of spacecraft by industry. Moreover, while an individual company in one of the major European countries could perhaps efficiently develop an entire mission, the political integrity of ESA relies on a system of 'juste retour' (fair returns) such that each member state receives contracts on a major project in the same proportion as its contribution to the ESA budget. Thus France and Germany must each be contracted to perform about 25% of the work, with Italy and Britain a little less and then a few per cent for Denmark, Switzerland, Sweden and so on. This constraint imposes a tremendous management overhead on how a space mission can be implemented.

The industrial team proposed by Aerospatiale (after some juggling of contracts and responsibilities to ensure an acceptable distribution) was responsible for the overall system design of the Huygens probe. Their affiliate, Aerospatiale of Les Mureaux near Paris, was to provide the all-important heatshield. The structure would be made by CASA of Spain, and the radio and computer system by Alenia Spazio and Laben of Italy. Germany's MBB would be responsible for the thermal design, and for the intensive program of system-level testing. The UK contribution included onboard software development by Logica, and the descent control system involving parachutes and pyrotechnics supplied by Martin-Baker – a company famous for its ejection seats.

Development of a new spacecraft usually entails the construction of several units of improving fidelity – not only are features of

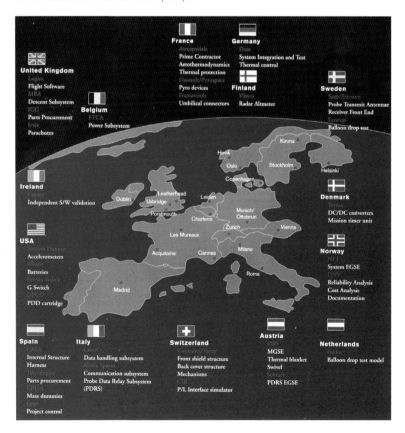

RIGHT The Structural, Thermal and Pyro Model of Huygens being readied for tests. Note the tiles on the front shield and bundles of cables at the right for test sensors. *(ESA))*

construction and operation often only found 'by doing', but it is usually necessary to test a system beyond what it will actually have to endure, and such testing could stress or weaken the flight unit. In Huygens's case four models were developed at system level:

■ A Structural, Thermal and Pyro Model (STPM) to qualify the probe design (including all of the mechanisms to be activated by pyrotechnic devices) for all structural, mechanical, and thermal requirements.

LEFT **The Engineering Model of Huygens at ESOC. It was essentially 'naked' equipment boxes mounted for ease of access with representative wire harness. The black boxes with cables mounted around the radar altimeter antennas on the periphery of the experiment platform were to provide simulated echo signals. The ACP and GCMS experiments can be seen behind the large green box of the power distribution unit. A HASI boom is visible in the foreground. Red covers protect the HASI temperature and pressure sensors. Tests were made with external power supplies – the batteries are not installed.** *(ESA)*

RIGHT The overall dimensions and configuration of Huygens. The Descent Module had a somewhat spherical lower section (fore dome) equipped with spin vanes, and a frustum upper section (aft cone). It sat inside the front shield and back cover. The probe was attached to the Cassini orbiter by a set of struts. *(Aerospatiale)*

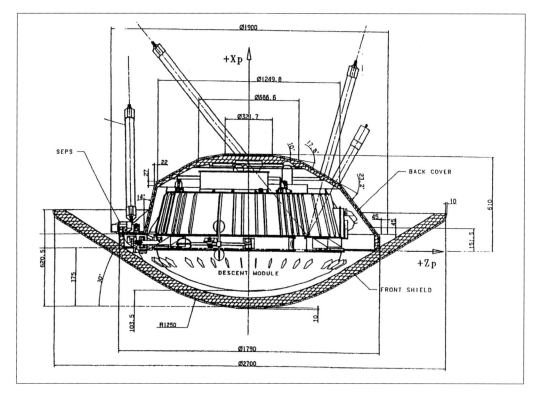

- An Electrical Model (more commonly referred to as an Engineering Model, EM) to verify the electrical performance of the probe, and also of its electrical and functional interfaces with the Cassini spacecraft.
- A Special Model (SM2) for a balloon drop test. All of the mechanisms and the descent control systems such as parachutes were flight-standard. (Concerning the numbering, it should be noted that an SM1 was initially planned but never built.)
- The Flight Model (FM) which would be delivered to Titan by Cassini.

Note that although there was no all-up Flight Spare (FS) probe, flight spare instruments were built (and in some cases were needed) in order to cope with failures during development.

The overall configuration of the Probe System comprised two principal elements:

- The 318kg Huygens probe that was to enter Titan's atmosphere after being released by the Saturn orbiter.
- The 30kg Probe Support Equipment (PSE), which would remain attached to the orbiter after the probe was released.

The Huygens probe consisted of the Entry Assembly (ENA) cocooning the Descent Module (DM). ENA provided orbiter attachment, umbilical separation and ejection, cruise and entry thermal protection, and entry deceleration control. It was jettisoned after entry, releasing the Descent Module, which consisted of an aluminium shell and inner structure containing all the experiments, and support subsystems including the parachute descent and spin control devices.

The PSE consisted of:

- Four electronic boxes aboard the orbiter: two Probe Support Avionics (PSA), the Receiver Front End (RFE) and the Receiver Ultra Stable Oscillator (RUSO).
- The Spin Eject Device (SED).
- The harness (including the umbilical connector) to provide power and RF and data links between the PSA, probe and orbiter.

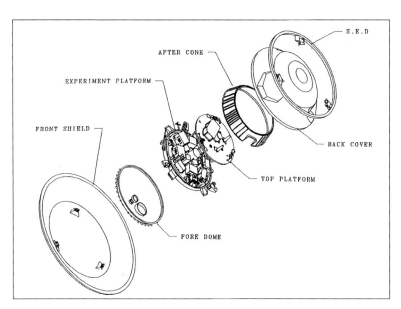

The **Front Shield Subsystem (FRSS)** needed to be large enough (2.7m) to decelerate the probe from the 6km/s entry speed to Mach 1.5 (~300m/s) above 160km altitude, where science measurements of Titan's stratosphere were desired. The kinetic energy of an object moving at 6km/s far exceeds the heat that will melt almost any material, and so the bulk of this

ABOVE The main subassemblies of the Huygens probe system.
(ESA/Aerospatiale)

BELOW The aerothermodynamic environment predicted for Huygens evolved as computational tools improved: after designing the heat shield against distressingly elevated heat fluxes in the early 1990s, revised calculations like these in 2004 suggested the system had very healthy margins *(EADS)*

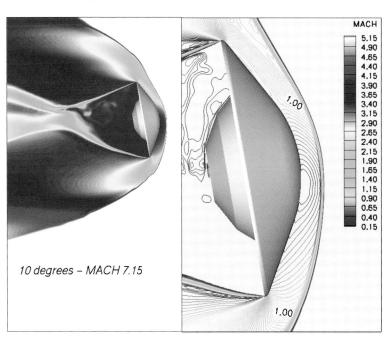

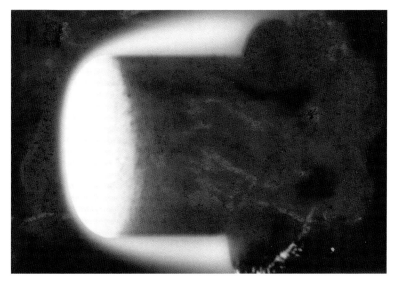

ABOVE The AQ60 thermal protection material was tested in a plasma wind tunnel, where a gas stream heated by an electric arc simulated the flow conditions of entry to assess how rapidly the heatshield material would melt, crack, or otherwise degrade under the punishing stress and heat flux.
(IRS Stuttgart)

energy must be deposited in the atmosphere, rather than in the object itself. The way to do this is to have a blunt-nosed object that develops a strong shock wave (the opposite tendency from supersonic aircraft, which are sharp-nosed to minimise drag). However, the vehicle needs to be sufficiently pointy to fly stably without tumbling, and so the optimum shape is a cone with a half-angle of about 60° with the cone blunted by a spherical nose.

The heat flux to be endured during entry would be about 1.4MW/m^2 (140W/cm^2), or

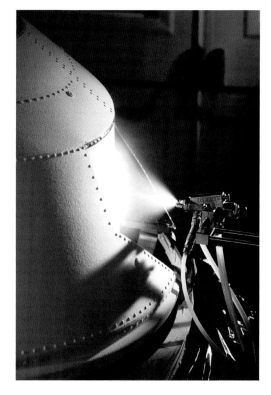

RIGHT Prosial insulation was sprayed onto the back cover of the Huygens probe.
(Airbus Defence and Space)

1,000 times that of full sunlight at Earth. To protect the probe from this load (which is comparable with Mars missions and Earth orbital capsules) the front of the shield would be covered in tiles of AQ60/I, a thermal protection system (TPS) material used on French ballistic missile warheads. AQ60/I is a porous (84%) felt of silica fibres (10%) reinforced by phenolic resin (6%). Between 200–1,000°C the resin pyrolyses and ablates away but leaves a stiff insulating residue, making it an effective TPS. The tiles were bonded to the 32kg FRSS structure, made of a Carbon Fibre Reinforced Plastic (CFRP) honeycomb sandwich, using 9kg of CAF/730 silicone adhesive. In order to tolerate the total heat load during the ~20sec entry heat pulse the tiles were ~18mm thick with an overall mass of ~30kg.

The **Back Cover Subsystem (BCSS)** had to protect the DM during entry. It carried multi-layer insulation (MLI) for the cruise and coast phases. Since it did not have stringent aerothermodynamic or structural requirements it was simply a stiffened aluminium shell of minimal mass (11.4kg) protected by Prosial (5kg). It included an access door for late integration and forced-air ground cooling of the probe, a breakout patch through which the initial (pilot) parachute would be fired, and a labyrinth seal with the front shield to provide a non-structural thermal and particulate barrier. To allow the probe to depressurise during launch the back cover had a vent hole.

The back side of Huygens would receive somewhat (~15 times) lower heat loads from the glowing shock that formed the luminous wake or 'meteor trail' of the probe than would the front shield. The composition of Titan's atmosphere meant these loads would not be insignificant, but a simpler TPS could be used – in this case a layer of Prosial, a suspension of hollow silica spheres in silicone elastomer, was sprayed on to the back cover and on to the aluminium facesheet of the front shield's aft surface.

The **Descent Control Subsystem (DCSS)** would control the descent rate in order to satisfy the payload's requirement for sufficient time at high altitude to make measurements, whilst keeping the total descent time to a

The stages of the descent control sequence, with the parachute and lines shown to scale. Note the long riser for the pilot parachute (since it must clear the wake of the large front shield) and the large size of the main parachute because it must pull the Descent Module safely out of the shield. *(ESA/Vorticity)*

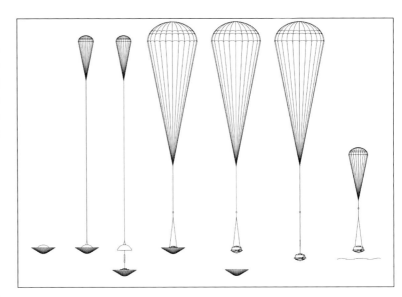

maximum of 2.5hr. The DCSS was activated nominally at Mach 1.5 and about 160km altitude. The sequence began by firing the Parachute Deployment Device (PDD), a mortar that ejected the pilot parachute pack. This was actuated by one of two NASA Standard Initiator (NSI) pyros which, using a transfer charge, ignited a carboxyl-terminated polybutadiene and ammonium perchlorate gas generator cartridge. The gas then blasted the parachute pack with a sabot (total ejected mass 1.28kg) through the back cover's breakout patch, in the process shearing its attachment pins.

The PDD actuation was timed to deploy the pilot parachute after the initial deceleration had slowed the probe to Mach 1.76 and at a dynamic pressure of less than 440 Pa. The 2.59m diameter Disk Gap Band (DGB) parachute was to inflate 27m behind the DM within 1.4sec. (The 27m is 10 forebody calibres, a rule of thumb to adequately reduce the wake effects of the probe on the supersonic parachute inflation.) Then 1.1sec later the back cover would be released, to be pulled away from the rest of the assembly by the pilot parachute.

As it receded, the back cover would pull a lanyard to extract the 8.30m diameter DGB main parachute from its container. This canopy was to inflate while the probe was still supersonic in order to decelerate and stabilise it passing through the transonic region (where the heatshield by itself would be unstable). Plain hemispheric parachutes do not have reliable inflation characteristics in this speed regime, but the annular gap in a DGB parachute stabilises the flow separation and provides much better characteristics. Using the same geometry for all three of Huygens's parachutes simplified the development and testing.

The front shield was to be released at about Mach 0.6, about 30sec after the inflation of the main chute, whose size was chosen to provide

BELOW The parachute deployment device. The Y-shaped section contains two redundant initiators to ignite the gas generator charge that blasted the pilot parachute and a sabot through a breakout patch in the back cover. At right is the cover, sabot, and pilot parachute bag (after the parachute had been extracted) during tests conducted in the UK. *(Author montage of images by S. Lingard)*

BOTTOM The Huygens parachute container. At left it is wired up with accelerometers (red cables) during a vibration test. At right the liner, a swivel, and some of the Kevlar lines are seen. The fabric-and-rope practicalities of parachute engineering provide an interesting contrast with the traditionally sterile realm of spacecraft development. *(Author montage of images by S. Lingard)*

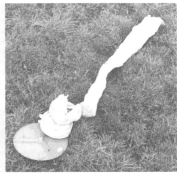

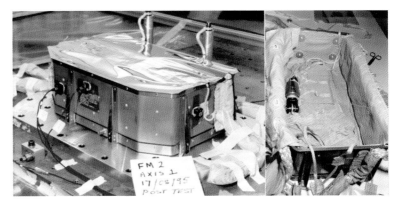

RIGHT The nylon Disk Gap Band parachute for Huygens during tests at Martin-Baker's facility in Chalgrove, England. The gap greatly increases the stability of the chute by controlling the flow separation. *(S. Lingard)*

RIGHT A parachute test vehicle, dropped by helicopter at Chalgrove under the stabiliser parachute (with the author in attendance). The bridle legs were 3.91m long and the canopy was trailed seven calibres behind the fore body. *(J. Underwood)*

sufficient deceleration to guarantee a positive separation of the front shield from the Descent Module.

The main parachute was too large for a nominal descent time shorter than 2.5hr, a constraint imposed by battery size limitations and the Cassini relay geometry, so it was to be released by the Parachute Jettison Mechanism (PJM) to permit the probe to fall faster under a 3.03m diameter DGB stabilising drogue parachute. The PJM comprised a bolt-cutter actuated by a 5A current pulse to one or both of two pyrotechnic initiators: the cutter would sever the 6.5mm stainless steel bolt which connected each leg of the main parachute bridle to the probe.

The main and stabiliser parachutes were housed in a single canister on the DM's top platform, protected by a thin aluminised Kapton cover. All three parachutes were made of nylon fabric using Kevlar lines. The lines to the canopy and the bridle lines were each attached to an aluminium ring at their confluence. The bridle was

BELOW LEFT Even as lowly a component as the parachute swivel required great thought into the loads it would have to sustain (in this case 14.7kN), and what materials could be used. Bellville springs maintained a preload on the bearing. *(S. Lingard)*

BELOW A parachute jettison mechanism (one of three). One or both of the two pyro bolt cutters (one is seen as the tube at the lower right) allowed clean separation of the clevis that held the spool around which the Kevlar parachute bridle was wrapped. Each such mechanism was attached to the top platform by four M8 bolts (not shown). *(S. Lingard)*

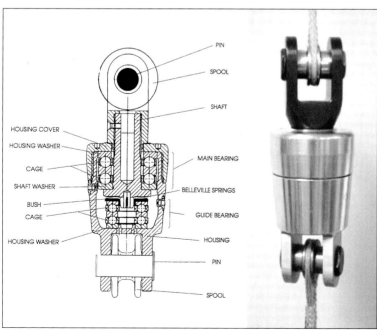

initially secured to the top platform with Velcro tie-downs bonded with Araldite AV138 adhesive. A Kevlar riser connected the two rings.

The spin of the probe was decoupled from that of the parachute by a swivel using redundant low-friction bearings in the connecting riser of both the main and stabiliser parachutes. The swivel used ball bearings with a molybdenum disulphide solid lubricant. There were two sets of bearings such that even if one set seized, the swivel would turn with minimal resistance torque.

The **Separation Subsystem (SEPS)** provided mechanical and electrical attachment to (and separation from) Cassini, and provided the transition between the entry configuration ('cocoon') and the descent configuration (DM under parachute).

The three SEPS mechanisms were connected on one side to the Inner Structure Subsystem of Huygens, and on the other side to Cassini's supporting struts. As well as forming the probe-orbiter structural load path, each SEPS fitting incorporated a pyronut for separating the probe from the orbiter, a rod cutter for front shield release, and a rod cutter for back cover release.

Within SEPS, the Spin Eject Device (SED) that performed the mechanical separation from the orbiter comprised the following elements:

- Three stainless steel springs delivered the separation force with a stroke of about 6cm, and were retained after actuation.
- Three guide tracks, each with two axial titanium rollers running along a 30° helical rail, ensured controlled ejection and spin (at 0.3m/s and 5–10rpm respectively). Their design ensured that the rollers only contacted the rails after they had already travelled 1–2mm, preventing any danger of cold-welding during cruise.
- A carbon fibre ring to accommodate the asymmetrical loads from the orbiter truss and to provide the necessary stiffness before and after separation.
- Three pyronuts to provide the mechanical link prior to separation.

In addition, the Umbilical Separation Mechanism (USM) which provided orbiter-probe electrical

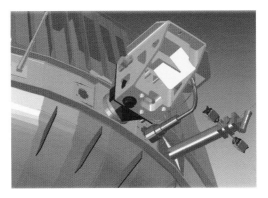

links via three 19-pin connectors, was disconnected by the SEDs. (Rather than using guillotine cutters, it had been decided that these connectors should be simply yanked apart by the momentum of separation.)

The **Inner Structure Subsystem (ISTS)** had mounting support for the probe's payload and subsystems. It was fully sealed except for a vent hole of about 6cm^2 on the top, which was there to prevent the pressure differential during ascent and descent from exceeding 5mbar. It comprised:

- The 73mm thick aluminium honeycomb sandwich experiment platform which supported most of the experiments and subsystems units, together with their associated harness.
- The 25mm thick aluminium honeycomb sandwich top platform that was the top external surface of the DM and supported

LEFT A computer-aided drawing (CAD) package view of the Descent Module, showing one of the three inner separation subsystem brackets. The retention of these brackets on the DM was not widely appreciated, and in fact they may have strongly influenced the spin of the probe as it descended on its parachute. At left is a discharge wick to prevent electrostatic charge build-up during the descent, and at right the Kiel probe for pressure sensor and two temperature sensors of the HASI instrument. *(Author)*

BELOW A cross-section of the DM structure. *(ESA/ Aerospatiale)*

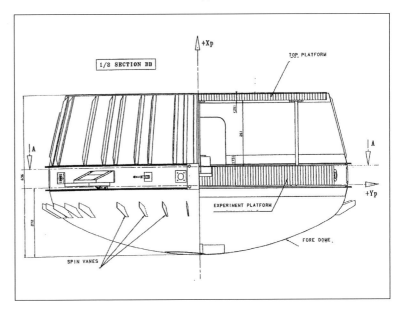

ABOVE **A view of the top platform of the Flight Model of the Huygens probe. At left is the cover over the DISR instrument. The cylinder at centre is the mortar for the pilot parachute, and the large foil-covered box adjacent to it holds the main and stabiliser parachutes as well as their lines and swivels. Three parachute jettison mechanisms can be seen around the periphery, and the two black cones are the probe transmit antennas. The CD-ROM for outreach is seen at roughly the one-o'clock position, and the spin-balance weights occupy the edge between about three-o'clock and six-o'clock.** *(DASA)*

BELOW **The experiment platform is barely visible during final checks at KSC, under the black electronics boxes, the harness, and the dozens of electrical connectors. Also visible is the black carbon fibre support ring that stiffened the mounting to the orbiter support struts. The flat radar altimeter antennas and SEPS mechanisms can be viewed around the perimeter. The HASI booms and temperature sensors are covered with red pre-flight covers.** *(NASA/KSC)*

the Descent Control Subsystem and probe RF antennas.

- The after cone and fore dome aluminium shells, linked by a central ring.
- Three radial titanium struts which interfaced with SEPS and ensured thermal decoupling.
- Three vertical titanium struts that linked the two platforms and transferred the main parachute deployment loads.
- 36 spin vanes on the fore dome's periphery to provide spin control during descent.
- The secondary structure for mounting experiments and equipment.

Late in integration, 2.85kg of spin balance weights were attached to tune the moments of inertia and centre of mass of the probe.

During the time that the PSE was thermally controlled by the orbiter, the probe's **Thermal Subsystem (THSS)** had to maintain all experiments and subsystem units within their viable temperature ranges. In space, the THSS partially insulated the probe from the orbiter and ensured only small variations in the probe's internal temperatures, despite the incident solar flux varying from 3,800W/m^2 (in the vicinity of Venus, where the energy in sunlight is about three times that at Earth) to 17W/m^2 (approaching Titan after a lengthy coast phase after orbiter separation with about 90 times less sunshine than is received by Earth).

Probe thermal control was achieved by:

- Blankets of multi-layer insulation (MLI) surrounding all external areas, with the exception of the small 'thermal window' of the front shield. MLI is usually several dozen layers of a thin (10μm or less) Mylar or Kapton film with a metallic coating to prevent radiative heat transfer in the vacuum of space (the orange Kapton often makes blankets look gold in colour). Sometimes a fine mesh or scrim or a texture of indentations is used to minimise the contact area between the sheets.
- 35 Radioisotope Heater Units (RHU) on the experiment platform and top platform continuously provided about 1W each, even when the probe was dormant.

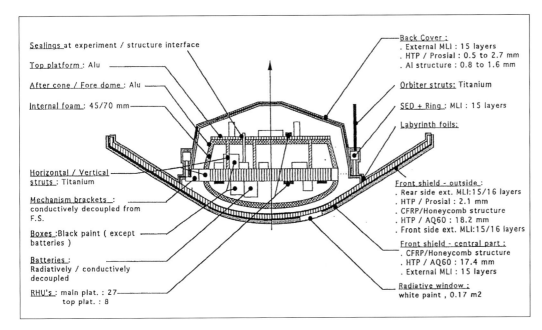

Sealings at experiment / structure interface

Top platform : Alu

After cone / Fore dome : Alu

Internal foam : 45/70 mm

Horizontal / Vertical
struts : Titanium

Mechanism brackets :
conductively decoupled from
F.S.

Boxes :Black paint (except
batteries)

Batteries :
Radiatively / conductively
decoupled

RHU's : main plat. : 27
          top plat. : 8

Back Cover :
. External MLI : 15 layers
. HTP / Prosial : 0.5 to 2.7 mm
. Al structure : 0.8 to 1.6 mm

Orbiter struts: Titanium

SED + Ring : MLI : 15 layers

Labyrinth foils:

Front shield - outside :
. Rear side ext. MLI:15/16 layers
. HTP / Prosial : 2.1 mm
. CFRP/Honeycomb structure
. HTP / AQ60 : 18.2 mm
. Front side ext. MLI:15/16 layers

Front shield - central part :
. CFRP/Honeycomb structure
. HTP / AQ60 : 17.4 mm
. External MLI : 15 layers

Radiative window :
white paint , 0.17 m2

■ A white 0.17m^2 thin aluminium sheet (the 'radiative window' of HINCOM/NS43G white paint) on the front shield's forward face acted as a controlled heat leak (about 8W during cruise) to reduce the sensitivity of thermal performance to variations in MLI efficiency (e.g. contamination).

The MLI would burn and be ripped away during entry, leaving temperature control to the AQ60 high-temperature tiles on the front shield's front face, and to Prosial on the front shield's aft surface and on the back cover.

During the descent phase, thermal control would be provided by foam insulation and gas-tight seals. A lightweight open-cell Basotect foam covered the internal walls of the DM's shell and top platform. This foam would prevent convection cooling by Titan's cold atmosphere (70K at 45km altitude) and thermally decouple the units mounted on the experiment platform from the cold aluminium shells. Gas-tight seals around all elements protruding through the DM's shell would minimise gas influx. Actually, the DM was gas tight except for the single 6cm^2 vent hole.

RIGHT Multilayer insulation blankets are attached onto the front shield during assembly in Germany. This view shows nicely the TPS tiles and white radiative window. (ESA)

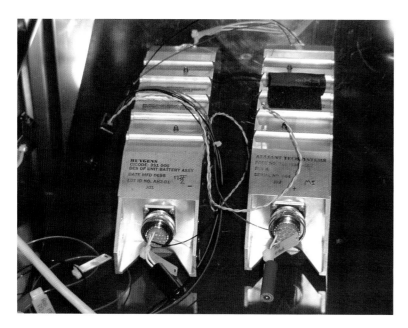

The *Electrical Power Subsystem (EPSS)* consisted of:

■ Five batteries to provide mission power from orbiter separation through to at least 30min after arrival on Titan's surface. Each battery comprised two modules of 13 lithium sulphur dioxide D-cells in series (G3108A3 cells made by Alliant). These cells were chosen because of their long storage life (they had been qualified for use on the Galileo probe) and high energy density. They have a nominal capacity of 7.6Ah. Each cell had a shunting diode to protect against open-circuit failure. The nominal cell voltage was 2.75V.

■ Power Conditioning and Distribution Unit (PCDU), the largest individual unit box on the probe, provided the power conditioning and distribution to the probe's equipment and experiments via a regulated main bus, with protection to ensure uninterrupted operations even in the event of single failure inside or outside the PCDU.

During the cruise phase, the Huygens probe was powered by the Cassini orbiter and the PCDU isolated the batteries. Five interface circuits connected to the Solid State Power Switches (SSPS) of the orbiter provided probe-orbiter insulation and voltage adaptation between the SSPS output and the input of the Battery Discharge Regulator (BDR) circuits of the PCDU. The BDRs conditioned the

power from either the orbiter or the batteries, accepting 32–78V and using a pulse-width modulated buck circuit to regulate the 28V bus to within 300mV. The power distribution was performed by 28 active current limiters which would disconnect any load in the event that it failed short-circuit, with the current limitation adapted for each user (e.g. the ACP heater limited at 2.7A). The Mission Timer, however, was supplied directly from the battery lines.

The Pyro Unit (PYRO) provided two redundant sets of 13 pyro lines, directly connected to the centre taps of two batteries (through protection devices), for activating pyro devices that were fired by current pulses lasting 50ms. Safety requirements were met by three independent levels of control relays in series in the Pyro Unit, as well as active switches and current limiters controlling the firing current. Three types were used:

■ Energy intercept relay, activated by PCDU at the end of the coast phase.
■ Arming relays, activated by the arming timer hardware.
■ Selection relays, activated by Command and Data Management Unit (CDMU) software.

In addition, safe/arm plugs were provided on the unit itself for ground operations – by shorting out pyro lines, these plugs prevented inadvertent firing before launch.

The EPSS operated in different modes depending on the mission phase:

■ Cruise phase: The EPSS was completely OFF over the whole cruise phase, except for the periodic checkout operations. There was no power at the interface to the orbiter, and direct monitoring by the orbiter allowed verification that all the relays were open.
■ Cruise phase checkout: The EPSS was powered by the orbiter for the cruise checkout operations. The 28V bus was regulated by the EPSS BDRs associated with each orbiter SSPS; a total of 210W was available from the orbiter and all the relays were open.
■ Timer loading: Following the loading (from the orbiter) of the correct coast time duration into the Mission Timer Unit, battery

depassivation (a high current draw to burn off an oxide layer on the electrodes) was performed in order to maximise energy output after the long cruise. Before probe separation, the EPSS timer relays were closed to supply the Mission Timer from the batteries and the orbiter power was switched off.

■ Coast phase: Only the Mission Timer was on, supplied by the batteries through specific timer relays during the coast phase. The EPSS was switched off and all other relays were open.

■ Probe wake-up: After counting down the coast phase, the Mission Timer awakened the probe by activating the EPSS. Input relays were closed and the current limiters powering the CDMU were automatically switched on as soon as the 28V bus reached its nominal value (other current limiters were initially off at power up). The pyro energy intercept relay was also automatically switched on by a command from the PCDU.

■ Entry and descent phases: All PCDU relays were closed and the total power (nominally 300W, at most 400W) was available on the 28V distribution outputs to subsystems and equipment. The Pyro Unit performed the selection and firing of the squibs in response to CDMU commands.

The data handling and processing functions were divided between the Probe Support Equipment (PSE) on the Cassini orbiter and the CDMUs – part of the **Command and Data Management Subsystems (CDMS)** – in the Huygens probe. The Probe Data Relay Subsystem (PDRS) provided the radio frequency (RF) link function for this purpose, together with the data handling and communication function with the orbiter's Control and Data Subsystem (CDS) via a Bus Interface Unit (BIU). During ground operations and cruise phase checkouts the orbiter-probe RF link was replaced by umbilical connections.

The CDMS had two primary functions: autonomous control of probe operations after separation; and management of data transfer from the equipment, subsystems, and experiments to the probe transmitter for relay to the orbiter. For these functions, the CDMS used the Probe Onboard Software (POSW), for which

it provided the necessary processing, storage and interface capabilities.

The driving requirement of the CDMS design was intrinsic single-point failure-tolerance. As a result of the highly specific Huygens mission (limited duration and no access by telecommand after separation), a very safe redundancy scheme was selected. The CDMS comprised:

■ Two identical CDMUs.
■ A triply redundant Mission Timer Unit (MTU).
■ Two mechanical g-switches (backing up the MTU).
■ A triply redundant Central Acceleration Sensor Unit (CASU).
■ Two sets of two mechanical g-switches (backing up the CASU).
■ A Radial Acceleration Sensor Unit (RASU) with two accelerometers.
■ Two Radar Altimeter proximity sensors, each comprising separate electronics, transmit antenna and receive antenna.

BELOW The command and data management system of the probe. The overriding design principle was redundancy for reliability with two essentially independent operating chains. *(ESA)*

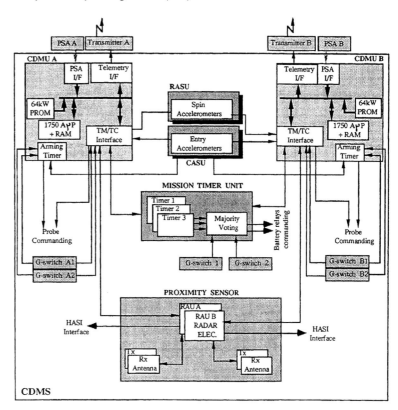

The two CDMUs each executed their own copies of POSW simultaneously and were configured with hot redundancy using Chain-A and Chain-B. Each hardware chain could run the mission independently. They were identical in almost all respects – the following minor differences facilitated simultaneous operations and capitalised on the redundancy:

- Telemetry was transmitted at two different radio frequencies.
- Chain-B telemetry was delayed by about 6sec to avoid loss of data should a temporary loss of the telemetry link occur (such as antenna misalignment as Huygens oscillated beneath its parachute).

Each CDMU chain incorporated a health check (called the Processor Valid status) which was reported to the experiments in the Descent Data Broadcasts (DDB). A chain declared itself invalid when it detected two bit errors in the same memory word, when an Ada exception occurred, or when an under-voltage on the 5V line occurred within the CDMU.

Each **Command and Data Management Unit (CDMU)** incorporated a 16-bit MAS-281 microprocessor running at 10MHz (a rather sedate pace compared with modern GHz processor speeds). The MAS-281, built by Marconi Electronic Devices, had a silicon-on-sapphire construction to harden it to radiation, and it implemented the MIL-STD-1750A architecture. The CDMU had a 64 kilo-word PROM for storing the POSW and a 64 kilo-word RAM to hold the POSW and other dynamic data when the CDMU was on (in flight only about half of the PROM was used and about 49K of the RAM). A Memory Management Unit was implemented to provide memory flexibility and some potential for growth. Direct Memory Access (DMA) was provided to facilitate data transfer between the memory and the input-output registers, thus relieving the microprocessor of repetitive input-output tasks. The RAM-stored program memory was protected against single-error occurrence by an Error Detection And Correction (EDAC) device, which could detect and correct single-bit errors and could

report any double-bit errors to the Processor Valid function.

In addition to conventional functions, the CDMUs had several Huygens-specific features:

- The arming timer function sent pyro and arming commands after a specific hardware-managed timeline, thereby offering full decoupling from the POSW operation.
- The Processor Valid signal was sent to experiments via the Descent Data Broadcast (DDB), to indicate the health of the nominal CDMU (Chain-A).
- Reprogrammability through the use of 16 kilowords of Electrically Erasable PROM (EEPROM) allowed patching of the POSW if necessary.
- The EDAC error count reported on internal data transfers.
- The capability to delay one telemetry chain via a specific RAM of 16 kilowords.

Each CDMU read 48 analogue channels, 14 thermistors, 40 digital lines, 16 relay status lines, and 10 serial channels.

The **Mission Timer Unit (MTU)** was used to activate Huygens at the end of the coast phase. To have a design tolerant to single-point failures, the MTU was based on three independent hot-redundant timer circuits and two hot-redundant command circuits. Two mechanical g-switches provided backup. MTU power was supplied directly via three 65V supply lines, one for each Timer Board, from independent batteries without going through the PCDU.

During the pre-separation programming activities, when Huygens was still connected to the orbiter, all three timers were loaded with the exact duration of the coast phase using serial memory load interfaces from one of the two CDMUs. Each of the three Timer Boards could be loaded independently from either CDMU. The programmed values could be verified by the serial telemetry channels. When programming was completed, the CDMUs and all other probe systems except the MTU were turned off and the probe was separated.

During the coast phase lasting about 22 days, the timer register was decremented by a very precise clock signal. The MTU consumed about 300mW during this period, since only the necessary circuits (CMOS-based) were powered on. Nominally, the probe would be awakened (with the CDMUs being powered on to initiate the sequence) when two of the three timers indicate the coast period had ended, making the mission tolerant to one timer failure. Should two timers fail, probe wake-up could be initiated (albeit somewhat late) once both g-switches triggered during entry.

When the Command Board majority voting detected either both g-switches active or at least two of the three 'time-out' signals received, five High Level Commands (HLC) were issued sequentially from each board to the PCDU in order to switch on both CDMUs. The timer then returned to a standby mode. Two g-switches ensured probe wake-up in the event of atmospheric entry without the time-out signal from any of the timers. These were purely mechanical devices that activated when deceleration imposed a load of 5.5–6.5g.

The **Central Acceleration Sensor Unit (CASU)** measured axial deceleration at the central point of the experiment platform during entry and the signal was processed by the CDMU to calculate the time for parachute deployment (t0). Its main building blocks were:

- The power circuit with two hot-redundant input power lines for single-point failure-tolerance.
- Three accelerometer analogue signal conditioning blocks. A low-pass filter with a 2Hz cut-off was used and the analogue output from each block was routed to both CDMUs. In addition, the design prevented failure propagation from one conditioning chain to the others.

The CASU read 0–10g (so the signal was actually 'clipped' during entry, a characteristic which was not important for this application). Back-up detection of t0 was performed separately for both CDMUs by two pairs of mechanical g-switches (42-gram masses held by a magnet, such that when the g-load exceeded the threshold, the mass fell off and pressed a microswitch) in case the primary

CASU were to fail. The threshold values for each pair of g-switches were 5.5g and 1.2g.

The **Radial Acceleration Sensor Unit (RASU)** measured the radial acceleration at the periphery of the experiment platform. This acceleration was assumed to be due to the centripetal effects caused by the probe spin. The signal was processed every 2sec by the CDMU in order to provide the spin rate of the probe for insertion into the DDB distributed to experiments. The RASU was designed to measure spin acceleration within 0–120mg using a 41.67V/g scale factor (the least significant bit in the telemetered value was 0.1rpm). The sensor itself was the same as the CASU but included only two accelerometers.

The **Radar Altimeter Unit (RAU)** proximity sensor had a pair of fully redundant Frequency Modulated Continuous Wave (FMCW) radar altimeters working at 15.4GHz and 15.8GHz (Ku-band) to measure the altitude of the probe above Titan's surface, starting from about 25km. Each of the four antennas (two per altimeter) was a planar slot radiator array that provided an antenna gain of 25dB with a symmetrical full beam width of 7.9°. A continuous signal that was modulated in frequency with a rising and falling ramp waveform was transmitted. The received signal had a similar form but was delayed by the propagation time, therefore the range to the target was proportional (with a linear frequency modulation ramp) to the instantaneous frequency shift between the transmitted and received signals. Received signal data were also provided to the Huygens Atmospheric Structure Instrument (HASI) in order to infer Titan's surface roughness and topography.

The probe used the following algorithm to estimate its altitude in real time. Initially, the value was a prediction – a 'Time-Altitude Table' (TAT) assuming a nominal descent. If both altimeters reported results within 10% of the current estimate, the average value was used. If the two readings were more than 10% off, then the value closest to the TAT was chosen. The altitude estimate was provided in the Descent Data Broadcast to the experiments to trigger measurement sequences below 10km.

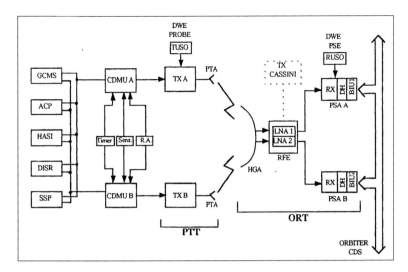

ABOVE **A block diagram illustrating the data paths from each of the six probe experiments via the two radio links, to the two receiving chains on the probe support avionics on the Cassini orbiter.** *(ESA)*

The ***Probe Data Relay Subsystem (PDRS)*** was Huygens's telecommunication subsystem, combining the functions of the RF link, data handling, and communications with the orbiter. It transmitted science and housekeeping data from Huygens to the PSE of the orbiter, which were then relayed to the orbiter CDS via a Bus Interface Unit. The PDRS was also responsible for telecommand distribution from the orbiter to the probe by umbilical during the ground and cruise checkouts. It comprised:

- Two hot-redundant S-band transmitters and two circularly polarised Probe Transmitting Antennas (PTA) on the probe.
- A Receiver Front End (RFE) unit (enclosing two Low Noise Amplifiers and a diplexer) and two Probe Support Avionics (PSA) units on the orbiter.

The orbiter's High-Gain Antenna (HGA) acted as the PDRS receive antenna. In addition, as part of the Doppler Wind Experiment (DWE), two ultra stable oscillators were available as reference signal sources for accurate measurement of the Doppler shift in the probe-orbiter RF link – with the Transmitter Ultra Stable Oscillator (TUSO) on Huygens and the Receiver Ultra Stable Oscillator (RUSO) on the orbiter.

The PDRS electrical architecture was identical on the two channels, except that

TUSO-RUSO were connected to only one of the two chains (A).

The ***Receiver Front End (RFE)*** of the Probe Support Equipment (PSE) comprised:

- A pair of Low Noise Amplifiers (LNA) linked to the orbiter's HGA to amplify the acquired RF signal by 20dB using two cascaded FET stages.
- Two RF inputs, one of which was linked to the HGA and the other via a coupler for use during checkout to link a dedicated transmitter output (on the probe) to the RFE via the umbilical.
- A pre-selection filter (coaxial cavity type with six poles).
- An isolator.
- An output attenuator (fixed value).

As a result of the HGA's shared use with the orbiter, a band pass filter (the TX filter) and a circulator protected the LNA Chain-B by isolating Cassini's S-band transmitter and Huygens's S-band receiver, making those two HGA modes mutually exclusive.

The two RFE outputs were sent to the two ***Probe Support Avionics (PSA)***, which performed detection, acquisition using a 256-point Fast Fourier Transform algorithm, tracking, signal demodulation, and data handling and management. The PSA data handling architecture was divided up into analogue and digital sections. The analogue section performed signal down-conversion from S-band to the IF frequency. The IF signal was quantised and the samples processed by the digital section.

The digital section performed:

- The Digital Signal Processing (DSP) function, specifically the signal acquisition and tracking task based on FFT analysis and frequency acquisition.
- The Viterbi decoding of the digital signal and delivery of the decoded transfer frame to the data handling section at 8,192 bit/s.
- The data handling task, which consisted of:
  - Transforming the received transfer frame into a telemetry packet.

- Generating internal PSA housekeeping data (including the synthesised frequency information) in a packet format.
- Controlling and managing communications with the orbiter CDS via a Bus Interface Unit (BIU).
- Distributing the telecommands from the orbiter BIU interface.

It comprised the following main modules:

- The receiver digital module, comprising the UT1750 microprocessor, 8,000 words of RAM and 8,000 words of PROM, and the receiver signal processing ASIC.
- The interface digital module, using gallium arsenide devices for Numerically Controlled Oscillator (NCO) and Digital to Analogue Converter (DAC) functions.
- The Support Interface Circuitry (SIC), which comprised the 8,000 words of EEPROM to memorise software patches; the 32,000 words of PROM containing the Support Avionics Software (SASW) and the testing, telecommand, telemetry and umbilical interfaces; and the MAS-281 microprocessor module used by the SASW.
- The BIU module that controlled communications between the PSA and the orbiter's 1553 bus.

The **Probe Transmitting Terminal (PTT)** comprised a pair of transmitters and two probe antennas. Each transmitter featured Temperature Controlled Crystal Oscillator (TCXO) synthesiser and BPSK modulator modules and a 10W Power Amplifier which used Automatic Level Control (ALC) for 40.2dBm nominal output power (end-of-life, worst-case, including ageing).

The reference oscillator for the Phase Locked Loop (PLL) synthesiser was either an (internal) Voltage Controlled Crystal Oscillator (VCXO) with a temperature compensating network or the (external) TUSO signal. The selection between these reference sources was made prior to separation from the orbiter. The TUSO had priority unless a failure was detected before separation.

The two transmitting antennas linked to the transmitters (dual chains without cross-coupling) employed quadrifilar helix designs.

The four spirals were fed at the bottom of the helix in phase quadrature. Left-hand circular polarisation (LHCP) provided signal transmission at 2,040MHz, and right-hand circular polarisation (RHCP) provided transmission at 2,098MHz. The minimum gain for the antennas, mounted on the top platform, was 0.9dB at all probe-orbiter aspect angles between +20° and +60°. The gain was highest towards the zenith, where Cassini was expected to be at the beginning of the descent, hence the antenna pattern partly compensated for the larger free-space loss at long range. At the end of the mission, when Cassini was expected to be closer, the system could afford lower antenna gain, although all these link budget calculations depended on how far east or west Huygens was blown by Titan's winds, which would not necessarily be known.

Software

Once, aerospace vehicles flew without software, but now software can demand a third or more of the development budget of a new system. Overall, Huygens's operations were rather simple. Indeed, in a sense it was not much more sophisticated than the clockwork timers that sequenced operations on the earliest rockets. Nevertheless, some events were reactive (such as the triggering of the parachute) and therefore needed to be reprogrammable.

Furthermore, in addition to the code running on the Huygens CDMS, known as POSW, and that in the PSA on the orbiter, known as the Support Avionics Software (SASW), the telemetry from POSW was relayed via the SASW and then Cassini's CDS to Earth. Two copies of the data handling hardware (i.e. CDMU and PSA) ran identical copies of POSW and SASW.

The software was based on a top-down hierarchical and modular approach using the Hierarchical Object-Oriented Design (HOOD) method and, except for some specific low level modules, was coded in Ada. The programming consisted, as much as possible, of a collation for synchronous processes timed by a hardware reference clock (8Hz repetition rate). The processes were designed to use data tables as much as possible. Mission profile reconfiguration and experiment polling could therefore be changed only by reprogramming these tables, which was possible via an EEPROM. To avoid the unnecessary complexity of a RAM modification while the software was running (with resulting unpredictable behaviour), direct RAM patching was forbidden. The POSW communicated with the SASW in different ways, depending on the mission phase.

Before probe separation, the two software subsystems communicated via an umbilical that provided both command and telemetry interfaces. Huygens couldn't be commanded after separation, and telemetry was transmitted to the orbiter via the PDRS RF link. The overall operational philosophy was that the software would run the nominal mission from power-up without checking its hardware environment or the probe's connection or disconnection. The specific software actions or inhibitions required during ground or flight check-out had therefore to be invoked using special procedures which were activated by the delivery of specific telecommands to the software.

To attain this autonomy, POSW's in-flight modification was autonomously applied at power-up using a non-volatile EEPROM. At power-up, the POSW would validate the CDMU EEPROM structure and then apply any software patches which were stored in the EEPROM prior to running the mission mode. If the EEPROM were found to be invalid at start-up, then no patches would be installed and execution would proceed employing the software in the CDMU ROM. A number of other checks were also carried out at start-up (e.g. a DMA check and a main ROM checksum), but the software would continue to attempt execution even if these start-up checks failed.

The **POSW** provided a number of functions.

Probe Mission Management:
- Detecting time t0 as entry began, based on the Central Accelerometer Sensor Unit signals.
- Forwarding commands at the correct times to the subsystems and experiments, according to the predefined mission timeline.
- Computing the spacecraft dynamical state from sensor readings.
- Sending Descent Data Broadcasts to the experiments.

Telemetry Management:
- Collecting and recording housekeeping data.
- Generating housekeeping packets from the housekeeping data.
- Collecting experiment packets according to a predefined polling scheme.
- Transmitting transfer frames to the PDRS.

Telecommand Management:
- Reception of telecommand packets from the PSE (while attached to the orbiter) following the ESA telecommand standard ESA PSS-04-107.
- Execution of commands related to these telecommand packets.
- Forwarding of commands to the experiments.

Control of the probe – the activation and forwarding of commands to experiments and subsystems – was driven by a predefined set of tables, the Mission Timeline Tables (MTT), that defined the actions which were to be performed as a function of time.

The pre-t0 MTT was activated when the probe was awakened from its cruise, and controlled it until the post-t0 MTT was activated by the POSW's detection of t0.

The experiments performed most of their activities autonomously based on the mission phase data which was computed within the POSW and sent to all the experiments every 2sec as a Data Descent Broadcast packet. The DDB contained the time, spin rate (computed by the POSW from the RASU signal or, in the event of failure, from a predefined look-up table) and altitude (initially taken from a look-up table based on the time elapsed since t0, but later by processing RAU data).

The telemetry management function involved the acquisition and transmission of probe telemetry in the standard packets (per ESA standard PSS-04-106; 16-bit words, most significant bit first). Whether they were housekeeping or experiment packets, they were all 126 bytes in length and were forwarded to the SASW in the form of transfer frames comprising header information followed by seven packets and then Reed-Solomon code words, making a total frame size of 1,024 bytes.

Housekeeping (HK) data were acquired from the subsystems (and from the software

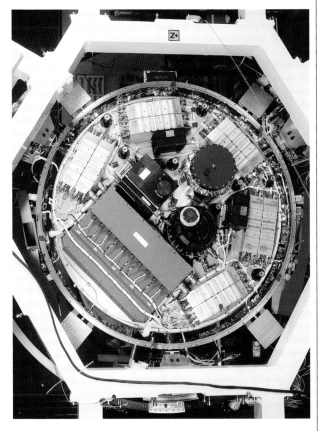
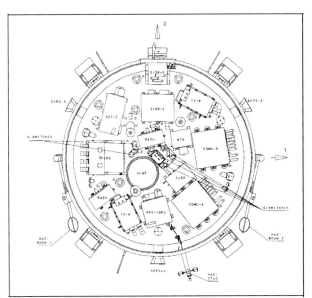
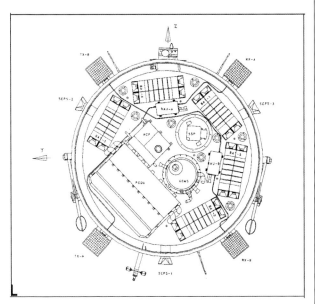

Accommodation of the payload and the major subsystems on the top/bottom of the experiment platform. ACP: Aerosol Collector Pyrolyser; BAT-I/S: batteries; CASU: Central Acceleration Sensor Unit; CDMU-A/B: Command and Data Management Unit; DISR: Descent Imager/Spectral Radiometer; DISR-E: DISR Electronics box; DISR-S: DISR Sensor Head; GCMS: Gas Chromatograph Mass Spectrometer; HASI: Huygens Atmospheric Structure Instrument; MTU: Mission Timer Unit; PCDU: Power Conditioning and Distribution Unit; PYRO: Pyro Unit; RASU: Radial Acceleration Sensor Unit; RUSO: Receiver Ultra Stable Oscillator; RX-A/B: receive antenna for Radar Altimeter A/B; SEPS: Separation Subsystem; SSP: Surface Science Package; SSP-E: SSP Electronics box; TUSO: Transmitter Ultra Stable Oscillator; TX-A/B: transmit antenna for radar altimeter A/B.

itself) at different rates according to a predetermined packet layout, and were loaded into four packets every 16sec. HK1 provided CDMU data and command counters; HK2 had PCDU data, temperatures and RASU readings; while HK3 carried the DDB, the RAU data and experiment ready flags. The HK4 packets with CASU data were buffered and issued 6.4min later as 'History', so that the data collected during entry were sent after the radio link was established.

Experiment data were acquired according to a predefined polling strategy, and the resulting packets were loaded into the transfer frames. The selection of an appropriate type of telemetry packet to include in each of the seven slots in a frame was managed by the polling sequence mechanism during a major acquisition cycle of 16sec (equal to 128 Computer Unit Times), driven by the Polling Sequence Table (PST) and the Experiment Polling Table (EPT). The PST defined whether housekeeping or experiment packets were to be included in the transfer frame that was currently under construction. However, it did not select which experiment was to be included. The EPT

defined the prioritised scheme for the collection of experiment data. The table was invoked whenever the PST requested experiment data for the transfer frame and was read in a cyclical manner. It consisted of a sequential list of the Huygens experiments, with the number of occurrences of each experiment in the table providing the polling priority.

By this method, the CDMS and the POSW were protected against failure modes in the experiments that could affect the data production rates. Each experiment was guaranteed an opportunity to supply data at, as a minimum, its nominal data rate. Furthermore, this polling scheme automatically optimised the data return by reallocating the TM resource in the absence of a 'packet ready' status flag from an experiment when expected.

Three EPTs provided different polling priorities during the descent's various stages, switching from one table to the next at a pre-set time.

The **SASW functions** were additional software provided by ESA for the PSA to manage the operation of the PDRS and transfer its telemetry and housekeeping data to the orbiter CDS via its MIL-STD-1553 bus using a BIU. While Huygens was attached, telecommands to update its software and conduct the in-flight checkouts were transmitted over the umbilical.

Positive protection

Could the probe be struck by lightning on Titan? The Voyager spacecraft did not detect any lightning flashes on the night-side of Titan, and its radio system did not detect any broadband electromagnetic 'spherics' or whistlers resulting from discharges there (although Saturn has strong emissions caused by lightning). However, not least because some of Huygens's instrumentation was capable of detecting lightning and thunder, the question was raised whether discharges in Titan's atmosphere might damage the Huygens probe or its instruments.

The Pioneer Venus probes suffered sensor failures for which some sort of electrical discharge was a suggested (albeit unlikely) cause, and instrumentation on Galileo had been designed with this in mind. Flying through charged droplets

BELOW The metal anti-static grille that was mounted in front of the 'Top Hat' cavity of the SSP to protect the instruments from possible electrical discharges. The acoustic sounder beamed its sound pulses through the grille. The circular hole was for the penetrometer. *(Author)*

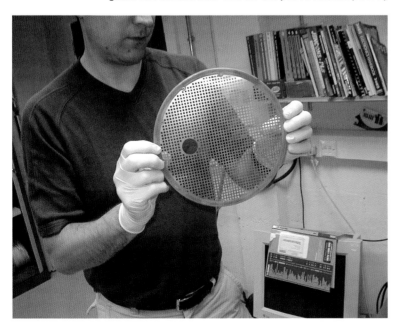

can lead to a build-up of electricity on an aircraft (which is why early aircraft often had metal tail skids instead of rubber wheels, to dissipate this charge on landing) and can sometimes cause displays of St Elmo's fire.

Liquid methane would provide a poorer dielectric for charge separation by cloud droplets than is water and the weak sunlight at Titan wouldn't drive much convective activity, so the probability of lightning seemed low, and if it did occur it would probably be weak. Nevertheless, ESA engineers (including the author) devised a specification based on that used for US aircraft and missile systems, but with a lower peak current (4kA instead of 20kA).

The probe was constructed as a conductive Faraday cage, protecting the electronics inside. Three discharge 'wicks' – fine wire points on the end of rigid stalks – were installed to prevent any charge build-up (similar wicks can be seen on the trailing edge of the wings of Airbus planes). Additionally, a lightning discharge test was made, zapping the probe engineering model with a massive current pulse. While the probe systems continued operating without problems, the ground support equipment rack in the adjacent room hung up and had to be rebooted!

Testing Huygens's instruments

Whilst Cassini and its instruments were designed for operation in the vacuum of space, and would have the opportunity to

ABOVE LEFT In an arrangement not dissimilar to the set of a Frankenstein movie, the probe was subjected to an electric discharge to simulate being struck by lightning on Titan. The probe is seen here upside down, with a large Marx generator (a ladder of giant capacitors and discharge gaps to generate the high-voltage impulse) in the background at the Universität der Bundeswehr in Munich, Germany. *(M. Hamelin)*

ABOVE The Engineering Model of the Huygens probe (encased in a somewhat simplified metal shell without spin vanes and stiffening ribs) was held upside-down beneath a ball-shaped electrode from which an intense electrical current pulse jumped for the lightning discharge test. One of the HASI booms is visible in its deployed position at centre. *(M. Hamelin)*

BELOW Inflating the balloon for a test of the HASI instrumentation in Sicily. The truck at right holds large cylinders of helium. *(F. Ferri)*

ABOVE Balloon tests always involve some uncertainty in recovery – but here the test article was safely tracked down to a field. The side of the box, made of insulating polystyrene foam, is visible at left, marked with a notice 'if found, please contact...'. *(M. Hamelin)*

ABOVE A mock-up of the Huygens probe used in a test of the HASI instrumentation and the radar altimeters. The large box that descended with the probe carried data acquisition and radio equipment. *(F.Ferri)*

BELOW An upward-looking test of the Huygens radar altimeters in the Netherlands. A box kite served as a point reflector in the sky. The reflections from nearby structures and the ground were suppressed by the blue pyramidal radar-absorbent material lining of the box around the probe. *(R. Trautner,/H. Svedhem/ESA)*

CASSINI-HUYGENS MANUAL

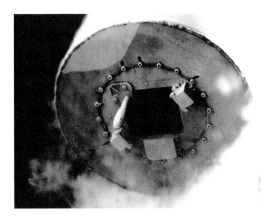

ABOVE Fog from liquid nitrogen coolant billows from the SSP 'Top Hat' assembly during a low-temperature test. *(SSP Team via J. Garry)*

RIGHT The Catalina mountains north of Tucson are visible in the background during this outdoor test of DISR on the roof of the University of Arizona's planetary sciences building. A light brown smear on the left of the upper housing underscores the fact that sometimes the test environment can be more hazardous than in flight – bird droppings were not encountered on Titan. *(C. See)*

make observations on the way to Saturn to understand their performance and to fine-tune operations, the Huygens probe's payload was destined for Titan's atmosphere and would only have one chance. The instruments were to be turned on once every six months for health checks, but because they were buttoned up inside the heatshield and in the vacuum and weightlessness of space during these tests there was relatively little that could be learned. Hence it was imperative to test the instruments on Earth in settings which were at least partly representative of descent on Titan, either by

RIGHT The DISR sensor head is visible at top left, mounted with batteries and a data acquisition rack on a tip/turn table on a tower atop Mount Bigelow, just north of the University of Arizona. This set-up allowed a sequence of real-world Sun sensor, image, and other data with an orientation history similar to what was expected from the descent of the probe, but in a form that was readily set up and modified. *(C. See)*

virtue of being up in the air, or by viewing useful scenes, or by measuring in similar conditions.

Testing Huygens's descent

Absolutely critical to Huygens's mission was the function of its parachutes, not only for stabilising and slowing its fall but also for extracting the Descent Module from the entry system.

Although stable inflation and steady-state drag performance can be measured in wind tunnels (and in this case a special large transonic wind tunnel in the USA had to be used because the tunnel needed to be large enough that the parachute would not 'feel' the constriction of the airflow at the tunnel walls) many of the functions of the Huygens Descent Control Subsystem were too rapid for steady state conditions to apply. So a free-flight test was needed.

What matters in a classic aerodynamic test is to match the Mach Number and Reynolds Number of the flow. Getting the Mach number right is important for understanding how the airflow compresses and how any shock waves may interact. The Reynolds number determines the effects of viscosity on the flow, and in particular how flow separates around objects to create a wake (golf balls are made rough to influence these effects). However, such a test can't replicate every feature of the Titan environment (especially as this wasn't known perfectly at that time).

There is an additional subtlety with parachutes in that the interaction of the deformable structure and the airflow goes in both directions. Therefore, matching the dynamic pressure is important especially for the mechanics of inflating a parachute. The Mach, Reynolds, and dynamic pressure conditions will change throughout a flight and it is usually impossible to match all three in a test on Earth. Consequently, some compromises typically have to be made.

RIGHT Triplets of images from the DISR helicopter flight. The upper panels were by the side-looking imager and they have a typical 'airplane window' view of the Arizona desert. The down-looking medium-resolution (middle) and high-resolution (bottom) images feature fields, gullies and trees. In near-infrared light (which dominates these images, because the instrument bandpass was optimised for Titan's hazy atmosphere) chlorophyll, and hence trees and healthy grass, appear bright. *(C.See/DISR Team)*

RIGHT The inflation and stability characteristics of the parachutes for the Huygens probe were tested at speeds in the range 350–1,000mph in a 16ft diameter transonic wind tunnel at the Arnold Engineering Development Center in Tennessee, USA. (US Air Force/AEDC)

For Huygens, the early parachute deployments and back cover and front shield separation mechanics were the most important items to verify. These would occur at an altitude of about 150km above the surface of Titan, where the ambient (static) pressure is 2mbar, about the same as the pressure at an altitude of 42km on Earth. This arises because Titan's low gravity 'stretches' the atmosphere vertically, in comparison with that of Earth.

Lifting the 320kg probe to 35km on Earth and allowing it to free-fall was determined to be the best option for Mach and dynamic pressure (the better speed match by some kind of rocket assistance was not worth the added complication and risk). The temperatures in Earth's stratosphere at this altitude would not be very different from Titan – the temperature influences the speed of sound and the viscosity of air, so the Reynolds and Mach values would be reasonable. The compromise in the balloon drop-test was that the pilot parachute deployment would be subsonic, but the main parachute inflation would be at the correct Mach and dynamic pressure (i.e. Mach 0.8 and 390Pa).

Only a few facilities worldwide could operate a helium balloon as large as that which would be needed for the Huygens test. Such balloons more usually carry telescopes into the stratosphere. In Europe, the ESRANGE facility near Kiruna in northern Sweden was selected as the most suitable, with a controlled area of 120x60km in which large objects could be safely

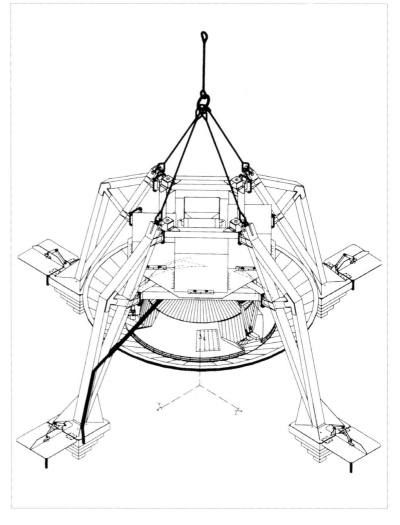

RIGHT Looking like an exotic planetary lander itself, an elaborate gondola was created to secure the Huygens probe for the SM2 stratospheric balloon test and to act as a data relay. Slabs of crushable material were placed on the footpads to limit any contact damage if the launch was not successful. (ESA)

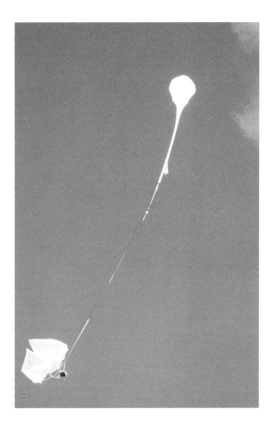

RIGHT The long teardrop shape of the SM2 balloon as it began its ascent. By the time the balloon reached an altitude of 37.4km, the gas would have inflated it to an almost spherical shape. Two ancillary tetrahedral balloons were attached to the gondola (released soon after take-off) to improve the safety of the launch process.
(ESA)

BELOW A sequence (not at equal intervals) of images from the upward-looking camera of the SM2 test high above arctic Sweden: (a) the balloon and gondola visible as two bright dots, after the probe was released into free-fall; (b) the mortar fires, blowing away the breakout patch as the riser to the pilot parachute snakes out behind the probe; (c) the inflated pilot parachute is visible through a window in the back cover, shortly after the pyros released the cover – the taut riser to the pilot parachute pulls the cover away; (d) the back cover recedes, pulling the main parachute bag at lower right; (e) the three-point bridle to the main parachute snakes upward as the parachute inflates; (f) the disk-gap-band main parachute is fully inflated and the bridle is taut; (g) after 10min the main bridle is released by pyro bolts and the probe falls from the parachute; (h) a line from the main bridle pulls out the bag that contains the stabiliser parachute, visible at the lower right; (i) the stabiliser inflates and pulls its own bridle taut.
(Author montage from an ESA video)

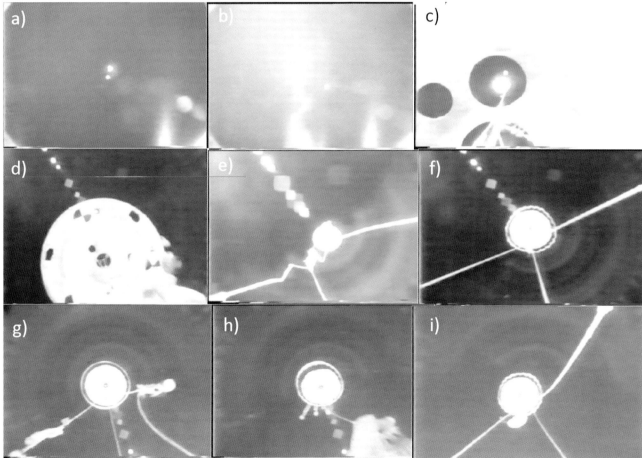

dropped from high altitude. The experience of the French space agency (CNES) in operating this type of balloon and the complicated communications and logistics of such tests was drawn upon, and their 402Z balloon was capable of lifting 1,000kg to 38km.

A special 'gondola' structure had to be designed and built to hold the probe securely when it was on the ground, and to control its release from the balloon at altitude. The probe to be dropped was a specially developed article (SM2) whose development was led by Fokker Space of the Netherlands; the structure was flight-like, complete with spin vanes and separation mechanisms, but without the probe instruments and electronics. Inside were gyros and accelerometers to measure the flight dynamics, as well as a GPS receiver. An S-band radio telemetry system was to transmit sensor data at 38kbit/s via the gondola, and the data stored onboard in a 16MB Solid State Recorder as backup. The separations of the parachute and front shield were to be recorded by upward-looking and downward-looking cine cameras.

The SM2 test was a major undertaking, with many international partners. After a Flight Acceptance review in April 1995, the systems were shipped to Sweden. If a major problem with the parachutes or separation system was discovered, this might seriously imperil the Huygens development schedule.

After a dress rehearsal the day before the planned text, and final weather observations on the day, the operators held their breath as the balloon began its 3hr ascent at 08:15hrs local time on 14 May 1995.

All the systems appeared to function during the 18min descent – the separation was clean, without any recontacts, and the parachutes inflated properly and didn't shred themselves

to pieces (as later happened in a NASA Mars parachute test). Some rapid oscillations of the probe were present in the gyro data, but these were attributed to wind conditions that were not expected to occur on Titan. The spin of the probe was not quite as fast as intended, so it was decided to increase the setting angle of the vanes on the flight probe from 2° to 3.2°. The direction in which the probe spun was not noticed, however.

In Earth's gravity and thinner air, the probe would land much more heavily than it would on Titan, so SM2 had an additional large cruciform recovery parachute to slow it down in the hope of minimising damage. In the event, it landed in snow and its recorders continued to operate. Apart from some scuffs and a couple of bent vanes, it was undamaged. It was a promising omen for the actual flight.

ABOVE The SM2 unit landed (with a special-purpose recovery parachute just seen at top left, not flown to Titan) in the arctic snow, gouging a dent and skidding slightly sideways (much as the real probe would do on Titan some eight years later). The probe was essentially undamaged and the instrumentation operated without interruption, raising confidence that the real Huygens probe might survive landing on Titan. *(ESA)*

BELOW Not every test was expensive. To gain familiarity with descent dynamics data, a University of Arizona team used a radio-controlled plane to release a small probe model with dataloggers and accelerometers from an altitude of 500ft on a model rocket parachute. *(Author)*

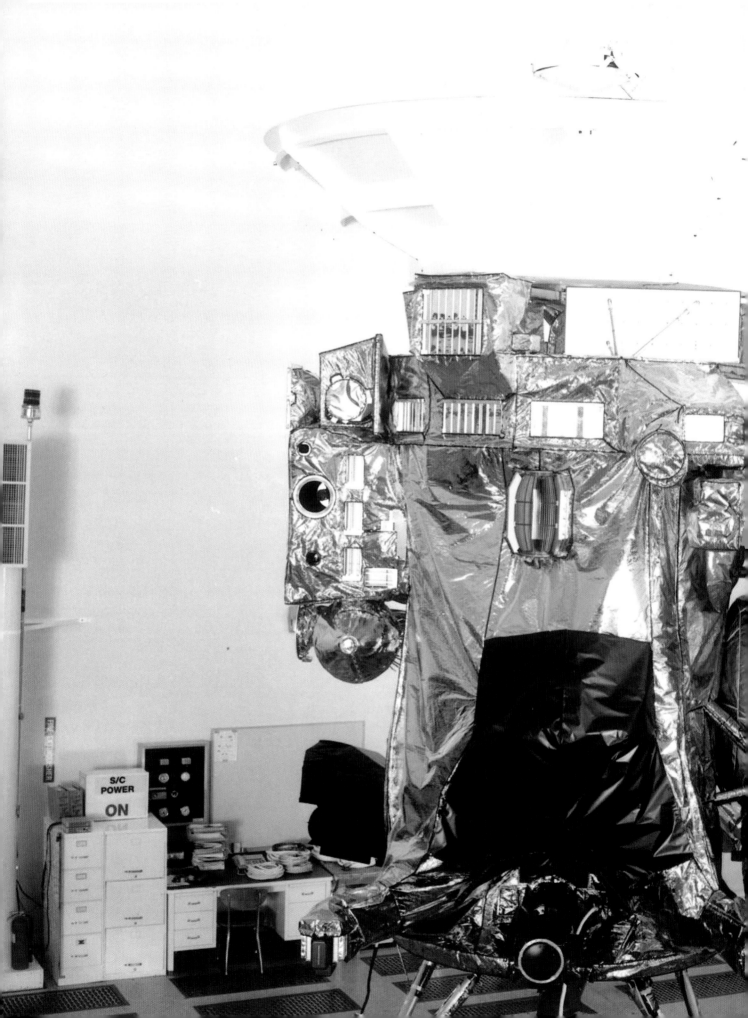

Chapter Three

The Cassini Saturn orbiter

The Cassini spacecraft was the most massive and complex planetary spacecraft flown by NASA. Its design had to be robust enough to ensure reliable operation near to and far from the Sun for over a decade. Cassini's massive arsenal of instruments would span huge ranges of wavelength and particle energies.

OPPOSITE The two-story tall Cassini spacecraft at JPL, most of its systems hidden behind gold or black thermal blankets *(NASA/JPL)*

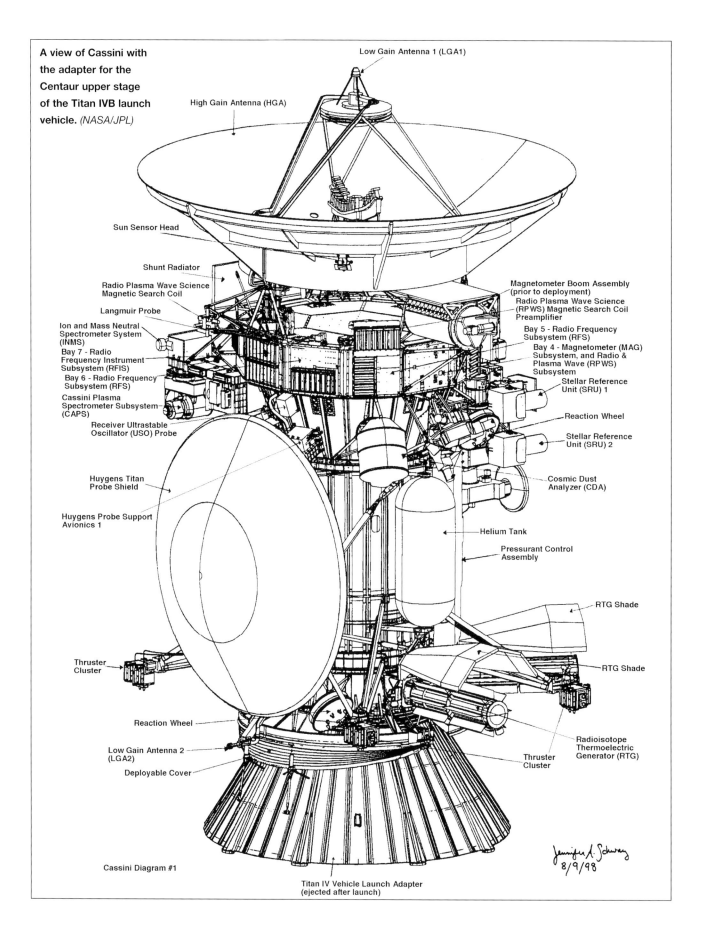

A view of Cassini with the adapter for the Centaur upper stage of the Titan IVB launch vehicle. *(NASA/JPL)*

Low Gain Antenna 1 (LGA1)

High Gain Antenna (HGA)

Sun Sensor Head

Shunt Radiator

Radio Plasma Wave Science Magnetic Search Coil

Langmuir Probe

Ion and Mass Neutral Spectrometer System (INMS)

Bay 7 - Radio Frequency Instrument Subsystem (RFIS)

Bay 6 - Radio Frequency Subsystem (RFS)

Cassini Plasma Spectrometer Subsystem (CAPS)

Receiver Ultrastable Oscillator (USO) Probe

Huygens Titan Probe Shield

Huygens Probe Support Avionics 1

Thruster Cluster

Reaction Wheel

Low Gain Antenna 2 (LGA2)

Deployable Cover

Magnetometer Boom Assembly (prior to deployment)

Radio Plasma Wave Science (RPWS) Magnetic Search Coil Preamplifier

Bay 5 - Radio Frequency Subsystem (RFS)

Bay 4 - Magnetometer (MAG) Subsystem, and Radio & Plasma Wave (RPWS) Subsystem

Stellar Reference Unit (SRU) 1

Reaction Wheel

Stellar Reference Unit (SRU) 2

Cosmic Dust Analyzer (CDA)

Helium Tank

Pressurant Control Assembly

RTG Shade

RTG Shade

Radioisotope Thermoelectric Generator (RTG)

Thruster Cluster

Cassini Diagram #1

Titan IV Vehicle Launch Adapter (ejected after launch)

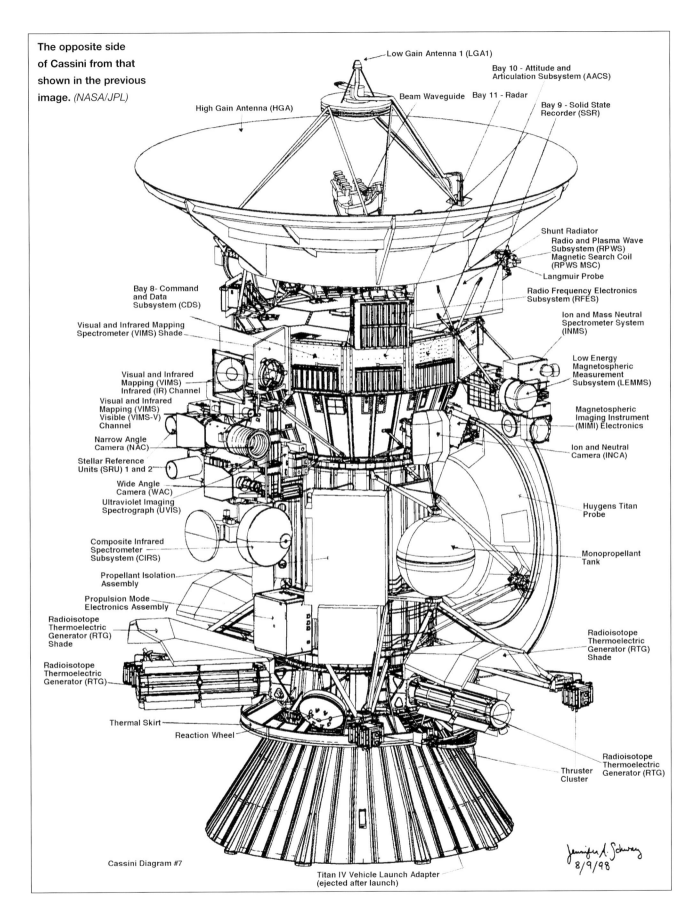

The opposite side of Cassini from that shown in the previous image. (NASA/JPL)

Low Gain Antenna 1 (LGA1)

Bay 10 - Attitude and Articulation Subsystem (AACS)

Beam Waveguide

Bay 11 - Radar

Bay 9 - Solid State Recorder (SSR)

High Gain Antenna (HGA)

Shunt Radiator

Radio and Plasma Wave Subsystem (RPWS) Magnetic Search Coil (RPWS MSC)

Langmuir Probe

Bay 8- Command and Data Subsystem (CDS)

Radio Frequency Electronics Subsystem (RFES)

Visual and Infrared Mapping Spectrometer (VIMS) Shade

Ion and Mass Neutral Spectrometer System (INMS)

Visual and Infrared Mapping (VIMS) Infrared (IR) Channel

Low Energy Magnetospheric Measurement Subsystem (LEMMS)

Visual and Infrared Mapping (VIMS) Visible (VIMS-V) Channel

Magnetospheric Imaging Instrument (MIMI) Electronics

Narrow Angle Camera (NAC)

Ion and Neutral Camera (INCA)

Stellar Reference Units (SRU) 1 and 2

Wide Angle Camera (WAC)

Ultraviolet Imaging Spectrograph (UVIS)

Huygens Titan Probe

Composite Infrared Spectrometer Subsystem (CIRS)

Monopropellant Tank

Propellant Isolation Assembly

Propulsion Mode Electronics Assembly

Radioisotope Thermoelectric Generator (RTG) Shade

Radioisotope Thermoelectric Generator (RTG) Shade

Radioisotope Thermoelectric Generator (RTG)

Thermal Skirt

Reaction Wheel

Radioisotope Thermoelectric Generator (RTG)

Thruster Cluster

Cassini Diagram #7

Jennifer A. Schwraz 8/9/98

Titan IV Vehicle Launch Adapter (ejected after launch)

Spacecraft design

The function of the Cassini spacecraft was to transport the Huygens probe to Titan and relay its data, to manoeuvre around the Saturnian system, and to support its science instruments by providing them with power and commanding and sending their data to Earth. The spacecraft also had to point these instruments and to some extent provide a benign environment (e.g. electrical grounding, some thermal control, etc.). But only once the actual payload was selected in 1990 could the detailed design proceed.

After the redesign of Cassini in 1992, the flight configuration with body-fixed instruments and no probe relay antenna converged. The spacecraft stood 6.8m (22.3ft) tall. The 4m High-Gain Antenna (HGA) could fully shield the rest of the spacecraft (except the deployed MAG boom and RPWS antennas) from sunlight when the HGA was pointed within 2.5° of the Sun. The HGA limit was set largely by the size of the fairing on the launch vehicle. The dry mass of the spacecraft was 2,523kg, including the Huygens probe system and the science

instruments (which added up to some 350kg). The best estimate of the actual spacecraft mass at separation from the Centaur upper stage was 5,573.8kg, of which about 3,050kg was propellant.

The main body of the spacecraft was formed by a stack consisting of the Lower Equipment Module, the Propulsion Module, the Upper Equipment Module, and the HGA. Attached to the stack was the frame supporting the Huygens probe, as well as the remote sensing pallet and the fields and particles pallet with their science instruments. Several instruments were installed separately: the magnetometers were on a 10.5m boom, and RPWS and CDA were fixed to the spacecraft body.

The Upper Equipment Module contained a 12-bay electronics compartment that including the Command and Data Subsystem and the Radio Frequency Subsystem. An additional 'penthouse' box accommodated the RADAR experiment.

The Lower Equipment Module carried the three Radioisotope Thermoelectric Generators (RTG) and the Reaction Wheel Assemblies (RWA) used for precision pointing of the spacecraft. Various electronics to support

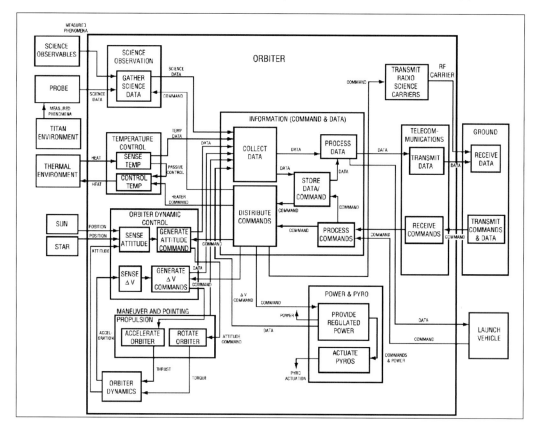

RIGHT Although this functional block diagram is purely schematic and mixes functions, hardware and other entities, it usefully depicts what a complex engineering system like Cassini must do. (NASA/JPL)

instruments and other spacecraft functions were also carried in the bus. During the inner cruise, the HGA and two Low-Gain Antennas (LGA) were used to transmit data and receive commands. One or other of the LGAs was selected when operational constraints prevented the HGA from pointing towards Earth.

The Cassini orbiter was a three-axis-stabilised spacecraft, meaning it typically pointed in a steady direction without spin (although it would occasionally be rolled around for scientific scans). The origin of the spacecraft coordinate system was located at the centre of the plane at the interface between the bus and the upper shell structure (i.e., the base of the electronic bays on the Upper Equipment Module).

The remote sensing pallet was mounted on the +X side of the spacecraft, the magnetometer boom was extended in the +Y direction, and the +Z axis completed the orthogonal body axes in the direction of the main engine. The primary remote sensing boresights viewed in the –Y direction, the probe was ejected in the –X direction, the HGA boresight was in the –Z direction, the main engine exhaust travelled in the +Z direction, and the main engine thrust was in the –Z direction.

Spacecraft subsystems

The spacecraft had a design life of 11 years (i.e. seven years of cruising plus a four-year mission at Saturn) and had to be designed with exceptional reliability. It comprised several subsystems, many of which were duplicated for redundancy, so that if one failed, the backup could be used. These included the computers and the main engine.

The **Structure Subsystem (STRU)** provided mechanical support and alignment for all flight equipment, including the Huygens probe. In addition it served as a local thermal reservoir and provided an electrical 'grounding', shielded against electromagnetic interference, and protected from radiation and meteoroids. The STRU consisted of the Upper Equipment Module (UEM) which contained the 12-bay electronics bus assembly, the instrument pallets and the MAG boom, and the Lower Equipment Module (LEM), plus all the brackets and

ABOVE Technicians gather around the Upper Equipment Module during launch assembly. It is dominated by the 4m diameter High-Gain Antenna. Note the 'umbrella' beneath the crane hook that was to prevent any debris from the roof of the building from falling onto the delicate space hardware. (NASA/KSC)

structure for integrating the Huygens probe, the HGA, LGAs, RTGs, reaction wheels, the main rocket engines, the four RCS thruster clusters, and other equipment. The STRU also included an adapter which supported the spacecraft on the Centaur during launch.

The **Radio Frequency Subsystem (RFS)** provided the telecommunications facilities for the orbiter and was also used as part of the radio science instrument. For telecommunications, it produced an X-band carrier at 8.4GHz, modulated it with data received from the Command and Data Subsystem (CDS) at a rate of up to 166kbit/s, amplified the X-band carrier power to produce 20W from the Travelling Wave Tube Amplifiers (TWTA), and delivered it to the Antenna Subsystem (ANT). From ANT, RFS accepted X-band ground command/data signals at 7.2GHz, demodulated them and supplied the commands/data to CDS for storage and/or execution.

The Ultra Stable Oscillator (USO), the Deep Space Transponder (DST), the X-band Travelling

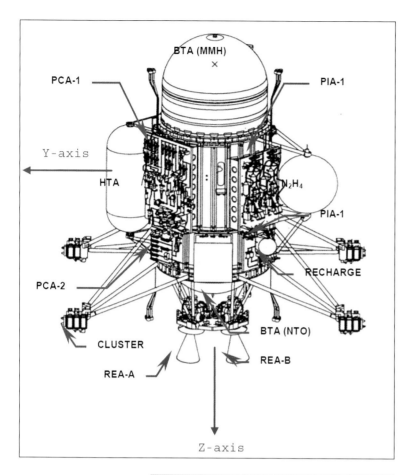

BTA (MMH)

PCA-1

PIA-1

Y-axis

HTA

N₂H₄

PIA-1

RECHARGE

PCA-2

CLUSTER

BTA (NTO)

REA-A

REA-B

Z-axis

Wave Tube Amplifier (TWTA), and the X-band Diplexer were elements of the Radio Frequency Subsystem (RFS) that were used as part of the radio science instrument. The DST could phase-lock to an X-band uplink and generate a coherent downlink carrier with a frequency translation adequate for transmission at X-, S-, or Ka-band. This two-way coherent link was used for precision Doppler tracking, for example to measure the gravity field of a satellite during a close flyby.

The DST could also accept the frequency reference signal from the USO in order to perform one-way measurements, for example in radio occultations. The DST had the capability of performing ranging measurements or modulating differential one-way ranging (DOR) tones onto the downlink to perform navigation.

The *Propulsion Module Subsystem (PMS)* provided thrust and torque to the spacecraft. As commanded by the Attitude and Articulation Control Subsystem (AACS), these thrusts and torques set the spacecraft attitude, pointing, and the amount of velocity change.

Originally, as in the case of Galileo, it was expected that the CRAF/Cassini propulsion systems would be provided by Germany, but upon the cancellation of CRAF this contribution was withdrawn, and NASA contracted Lockheed Martin Astronautics of Denver to build the system for Cassini.

For attitude control, the PMS had a Reaction Control Subsystem (RCS) consisting of four thruster clusters that were mounted off the PMS core structure, adjacent to the LEM at the base of the spacecraft. Each of these clusters contained four Olin hydrazine thrusters delivering a nominal thrust of 1N (equivalent to the weight of an apple in Earth's gravity). The thrusters were oriented to deliver their thrust along the ±Y and −Z axes of the spacecraft. They could also be used to provide delta-V (velocity change) to carry out small manoeuvres.

For larger delta-Vs, the PMS had a primary and redundant pressure-regulated main rocket engine. Each engine was a Kaiser-Marquardt R4-D motor producing a thrust of 445N when regulated (the R4-D is evolved, with a bigger, more efficient nozzle, from the attitude thrusters used by the Apollo spacecraft). The bipropellant

ABOVE A close-up of one of the '1-Newton' thrusters of the Cassini spacecraft. Catalyst degradation and the declining pressure of the hydrazine propellant as a result of use meant that later in the mission they produced only 0.7N. A fast-action solenoid valve injected hydrazine onto a heated catalyst where it violently decomposed into hydrogen, nitrogen and ammonia gases that were ejected through the tiny nozzle at lower right (inside the conical radiation shield). *(NASA/JPL-Caltech)*

ABOVE A close-up of the main engines with their special coating. The hottest section was at the top, in the combustion chamber and the narrow nozzle throat. The bowl-shaped radiation shield prevented heat damage to nearby structures from the red-hot glow. *(NASA/JPL-Caltech)*

main engines burned nitrogen tetroxide (N_2O_4) and monomethylhydrazine ($N_2H_3CH_3$). These are hypergolic propellants which ignite on contact. The engine was expected to provide a specific impulse of 308sec. These engines were gimballed, so when under AACS control during burns the thrust vector could be maintained through the shifting centre of mass of the spacecraft. AACS-provided valve drivers for all the engines and thrusters operated in response to commands received from AACS via the CDS data bus.

The PMS also included a retractable cover, like the roof of a convertible car, over the Main Engine Assembly (MEA) to shield it against micrometeoroids. This was because the main engine nozzles were considered a particularly vulnerable component, since they had a special disilicide coating to manage the strong heat load during engine firing, and if this coating were to become chipped, a hot spot might develop and allow burn-through.

The cylindrical bipropellant tanks had a diameter of 49in and were made from 6Al-4V titanium alloy, with a nominal volume of 49ft³. The maximum operating pressure was 330psi. The tanks used a set of eight metal vanes welded to the inside of the tank as propellant

BELOW The main engines with the cover in its retracted position. The two nozzles can be seen with red pre-flight covers. In the event of a motor failure, the cover could be jettisoned using pyro bolt cutters. By 2008 the cover (originally rated for 50 actuations, including 30 tests on the ground) had been operated 36 times in flight. Like most of Cassini's instruments and systems it was operated long after its warranty expired, but with redundant motors and careful monitoring and use, it was expected to complete 86 actuations in space by the conclusion of the mission in 2017. *(NASA/JPL-Caltech)*

RIGHT The stages in deploying the 2.1m diameter main engine cover, in this case using an engineering model. Only a thin blanket was needed to provide meteoroid shielding. *(NASA/JPL)*

management devices. These ensured that the slug of liquid propellant covered the outlet pipe from the tank in weightlessness, in order to maintain bubble-free flow of fluid. The 28in spherical monopropellant hydrazine tank, also 6Al-4V, had a maximum rated pressure of 420psi. The tank pressure would not be regulated and would 'blow down' as the hydrazine was used up, although it could be boosted one time during the mission from a small recharge tank (0.23ft^3, 3,600psi). The hydrazine tank used an elastomeric bladder for propellant management.

The health of the propulsion system was monitored by 81 temperature sensors, 18 pressure transducers, and 20 latch valve position indicators.

All in all, the system weighed 492kg dry and carried 8.7kg of helium, 132kg of hydrazine, 1,132kg of MMH and 1,868kg of NTO.

The ***Power and Pyrotechnics Subsystem (PPS)*** provided regulated electrical power from three RTGs on command from CDS to spacecraft users at 30VDC, distributed over a power bus. Both the high and low (return) rails of the bus were isolated from the chassis of the spacecraft by a resistance of 2,000 Ohms. Excess power was dissipated by a resistance shunt radiator mounted on top of the electronics bay and radiated to space.

Measurements of the output of the RTGs indicated a combined beginning-of-life power of 876 ±6W that was expected to fall to 740W upon arrival at Saturn and to 692W at the end of the prime mission. These estimates were at least 30W above pre-launch predictions. The generators each used the heat which was liberated by the decay of 10.9kg of plutonium

dioxide to produce about 300W of electrical power at the beginning of the mission. In addition to the decay of the plutonium itself, the power output would decline due to radiation damage to, and slow degradation of the silicon-germanium thermoelectric converters. At the very end of the mission in 2017 the output had degraded to just above 600W.

Electrical loads were connected to the power bus by 192 Solid State Power Switches (SSPS) that featured controlled ramp-up of voltage to loads and hardware controlled automatic shutoff in the event of over-current. Each switch could deliver up to 3A of current. Monitors of the on/off state, trip state, and load current through each switch were available in telemetry.

The PPS also supplied high instantaneous power to activate pyrotechnics. Because the RTGs provided continuous power (unlike solar cells whose output can be interrupted by turning away from the Sun, or by eclipses), Cassini did not require a battery, so the PPS incorporated two block redundant pyro switching units consisting of capacitor banks and associated electronics to provide the high current pulses for pyro firing (the RADAR instrument also had a capacitor bank to provide energy storage for the brief but very high instantaneous power needed to transmit pulses).

The PPS control electronics included hardware to detect and recover the spacecraft from system under-voltages. If the bus voltage were to drop below a preset threshold, the switches to many loads would be automatically switched off.

The ***Command and Data Subsystem (CDS)*** handled the reception of radio commands from Earth, their execution, and the routing of telemetry information from the various spacecraft systems and instruments. These were all linked together with a MIL-STD-1553B data bus which could robustly handle large data rates (e.g. greater than 400kbit/s from the RADAR experiment).

CDS software contained algorithms that provided protection for the spacecraft and the mission in the event of a fault. Fault protection software ensured that, in the case of a serious problem, the spacecraft would be placed into a safe, stable, commandable state (without ground intervention) for a period of at least two weeks in order to give the mission operations team time to solve the problem and send up a new command sequence. It was also capable of autonomously responding to a predefined set of faults which required immediate action. CDS itself comprised two 16-bit radiation-hard computers built to the MIL-STD-1750A standard and coded in Ada. Each computer had 8.2Mbits of RAM, 131kbits of PROM, and ran at a rate of 1.28 million instructions/sec.

The Cassini spacecraft included two identical Solid State Recorders (SSR). Each CDS (A and B) was attached to the two SSRs such that each CDS could read or write only one SSR at any one time. The Mission and Science Operations Office had the capability to control how the SSR attachments were configured via immediate command or a stored sequence. Under fault response conditions the

Flight Software (FSW) could switch an SSR attachment from CDS A to CDS B.

CDS received data from other onboard subsystems via the data bus, then processed and formatted them for telemetry and delivered them to RFS for transmission to Earth. Each subsystem interfaced with the MIL-STD-1553B data bus through a standard Bus Interface Unit (BIU) or a Remote Engineering Unit (REU). Data were collected in 8,800-bit frames, and Reed-Solomon encoded for downlink. Along with the encoding, a 32-bit frame-synchronisation marker increased these frames to 10,112 bits.

The **Attitude and Articulation Control Subsystem (AACS)** provided dynamic control of the spacecraft pointing, using its own redundant MIL-STD-1750A processors. In general, AACS was tasked to align spacecraft body vectors (e.g. the Z-axis engines, or an instrument boresight) along some inertial vector (e.g. the direction to a star, or some point on the surface of a moon). Two vectors had to be specified, a primary which the spacecraft could satisfy exactly, and a secondary, to which in general the spacecraft would get as close as it could (since the two might not be orthogonal).

Cassini maintained onboard a set of representations of the trajectories of different objects (e.g. the Sun, Earth, Titan, the F-ring and of course itself) typically as seven numbers known as orbital elements or as 12th-order

BELOW The Cassini AACS operated a control loop which took into account models of the spacecraft dynamics to accurately follow a given attitude profile. This example shows the loop architecture for control using the reaction wheels. *(NASA/JPL-Caltech)*

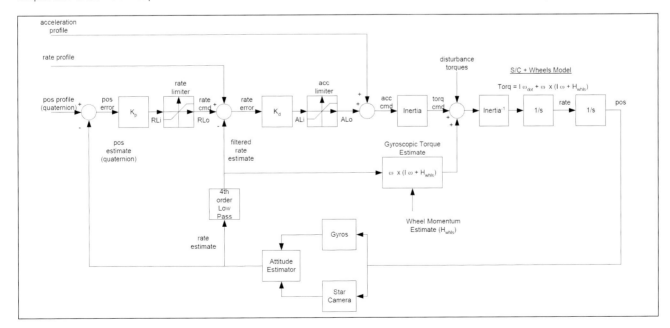

Cassini AACS Stellar Reference Unit S/N FM001

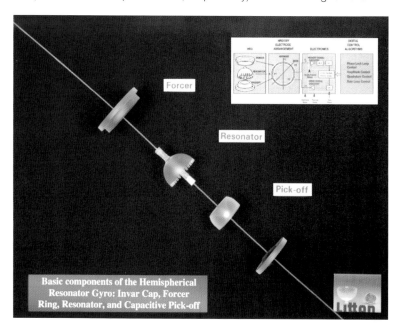

Basic components of the Hemispherical Resonator Gyro: Invar Cap, Forcer Ring, Resonator, and Capacitive Pick-off

RIGHT One of Cassini's star trackers. Note the lightweight ribbed aluminium structure and the dark baffles in the camera tube. Red covers are seen at lower left (probably over a reflecting cube for optical alignment) and above on a radiator plate which cooled the detector. *(JPL/NASA-Caltech)*

BELOW The core of the IRU was a 'wine glass' resonator. The structural pattern of the way that this 'sang' in response to vibration forcing was fixed in inertial space and could be sensed by pick-off electrodes, thereby serving as a gyro-compass in space. *(NASA/JPL-Caltech)*

Chebyshev polynomials from which positions could be calculated at a given time. Knowing these positions, the directions from the spacecraft could be computed by the Inertial Vector Propagator (IVP). This high degree of autonomy simplified the design of observations, and in particular permitted straightforward updating of the command sequences in response to new ephemeris information or a changed timeline.

On more primitive spacecraft, ground controllers would perform all the calculations and upload a low-level sequence of thruster commands or suchlike, and then *everything* would need to be revised if there were a small change.

On Cassini, the attitude instructions were of the form 'point the –X axis at Titan's north pole, with the +Y axis as close as possible to the Sun line (these being the primary and secondary vectors, respectively). In addition higher-level

attitude commands could perform repetitive subroutines, such as scans and mosaics.

AACS contained a suite of sensors that included redundant Sun Sensor Assemblies (SSA), redundant Stellar Reference Units (SRU) also called star trackers, a Z-axis accelerometer, and two three-axis 'gyro' Inertial Reference Units (IRU).

Each SRU viewed a field of view of 15x15° and could detect stars down to a visual magnitude of 5.6. It would compare the star field image with a catalogue of 5,000 stars to determine the spacecraft pointing to within about 1mrad (~0.06°) for precision scientific pointing. This star update was passed every 1–5sec to the attitude estimation logic, which used an Extended Kalman-Bucy filter also folded in data provided by the Inertial Reference Unit (IRU).

The IRU (of which, like the other critical units, there were two for redundancy) was an assembly of four Hemispherical Resonator Gyroscopes (HRG). These were not gyroscopes in the conventional sense of a spinning wheel (such mechanical gyros were often the life-limiting component on a spacecraft), but they used resonant vibrations in an axisymmetric quartz shell rather like the singing of a wine glass, to sense rotational motion. Three of the HRGs within an IRU were arranged orthogonally, and one diagonally to the others to serve as a consistency check.

These solid state sensors, now not uncommon, were new to space when Cassini was designed, and their bold choice proved judicious because the IRUs operated flawlessly (by 2013, IRU-A had been running for 137,000hr; the B unit was only powered on for occasional checks and to serve as a 'hot' backup for critical events).

The IRUs were crucial at times when the star trackers were blinded by bright objects in the field of view. In deep space, this could usually be avoided – indeed the Sun sensors were orthogonal to the SRU – but during close flybys of moons, Saturn or its rings large parts of the sky were filled by these bodies and the Star Identification (SID) algorithm was suspended in order to avoid confusing readings. During over 500 of these SID-suspend events lasting 4hr or more, the IRU data limited the pointing errors to a worst case of 50mrad (~3°); and since a scale factor update was uploaded in 2007, to below about 10mrad.

The SSAs, mainly used for reliable Sun-pointing or Earth-pointing for communications or safing recovery for which coarse pointing was adequate, had a field of view of 60° and provided an accuracy of 26mrad (a little over 1°).

Rotational motion during the Saturn tour that required high pointing stability was normally controlled by the three main Reaction Wheel Assemblies (RWA), although modes requiring faster rates or accelerations would use thrusters. Spinning a wheel faster would exert a torque back on the spacecraft in the opposite direction. An additional fourth reaction wheel could articulate to replace any single failed unit. Each unit (14kg, 159mm high and 369mm in diameter) had a 34Nms momentum storage capability and a motor capable of 0.165Nm of torque. An additional factor which limited the capability of the wheels was the allocated power of 90W for the set.

The **Temperature Control Subsystem (TEMP)** allowed operations over the expected solar ranges (0.61 to 10.1 AU) with some operational constraints. Temperatures of the various parts of the spacecraft were kept within allowable limits by a large number of local TEMP thermal control techniques, many of which were passive.

The 12-bay electronics bus had automatically positioned reflective louvres. Radioisotope Heater Units (RHU) were used where constant heat input rates were required and where radiation was not a problem. Multi-layer insulation blankets covered much of the spacecraft and its equipment.

Electric heaters were used in a variety of locations and operated by CDS and instruments. Temperature sensors were located at many positions on the spacecraft, and their measurements were used by CDS to command the TEMP heaters. Shading was executed by pointing the HGA (–Z axis) towards the Sun; the HGA was large enough to provide shade for the entire spacecraft body including the Huygens probe, but not the MAG boom or the RPWS antennas.

The Mechanical Devices Subsystem (MDS) provided the pyrotechnic separation device to disconnect the spacecraft from the launch vehicle

ABOVE Two of the RWAs (looking like flying saucers) are visible on the Lower Equipment Unit during assembly. (NASA/KSC)

BELOW Lights reflect off the slats of thermal control louvres just beneath the HGA in this assembly view. These plates automatically opened and closed to expose surfaces of different emissivity in order to passively regulate the temperatures of the electronics boxes. The thermal blankets have been pulled down to expose the structure and components underneath. (NASA/JPL)

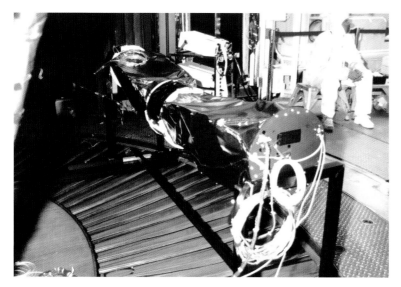

ABOVE The partly deployed MAG boom on a trestle fixture during thermal-vacuum testing at JPL. The person at right attests to the fact that 'thermal-vac' tests involve a great deal of sitting around. (NASA/JPL-Caltech)

adapter, with springs providing the impulse to push the spacecraft free. The MDS also provided a self-deploying 10.5m coiled longeron mast (the 'MAG boom') stored in a canister to carry the two magnetometers; electrostatic discharge covers over in-flight separation connectors; an articulation system for the backup reaction wheel assembly; a 'pin-puller' for the RPWS Langmuir probe; and louvres and variable RHUs for temperature control.

Because the MAG boom could not support

BELOW Rarely seen in its naked condition without the thermal blanket (undraped on the far side), the MAG boom assembly had a triangular set of elastic glass fibre longerons which spring out from a canister and were held rigid by a lattice of elements which become taut. (NASA/JPL-Caltech)

its weight sideways on Earth, for deployment tests, the boom had to be hung from a 'curtain-rail' arrangement, with ball-bearing trolleys sliding along a rigid track. Because the longerons (that define the three long edges of the triangular boom) were coiled up in their canister, the suspension fixtures were circular with bearings to accommodate rotation of the boom during deployment.

The **Electronic Packaging Subsystem (EPS)** consisted of the electronics packaging for most of the spacecraft in the form of the 12-bay electronics bus that was made of standardised, dual-shear plate aluminium electronics modules.

The **Solid State Recorder Subsystem (SSRS)** was the primary memory storage and retrieval devices on the Cassini orbiter. Solid state storage was expected to be more reliable and flexible than the magnetic tape recorders used by Voyager and Galileo, and this proved to be the case.

Each of the two Solid State Recorders (SSR) incorporated 128 submodules of Dynamic Random Access Memory (DRAM) of which 8 were used for Flight Software and 120 were used for engineering telemetry and science data. Each submodule could hold 16Mbit (16,777,200 bits) of data, so the total data storage on each SSR was 2.013Gbits. This storage capacity would set the limit on how much science data Cassini could acquire during each of its flybys.

720x4Mbit DRAMs, supplied by Oki Electric Industry Co., were installed in each SSR. Each recorder actually used only 640 of the DRAMs; the other 80 were spares that could be activated if some of the 'base' devices failed in-flight. Three large custom radiation-hard Application-Specific Integrated Circuits (ASIC) on each of six memory boards controlled access, refreshing, and error scrubbing for 40 DRAMs each. The DRAMs could withstand a total radiation dose of greater than 50,000 rad(Si) but were sensitive to Single-Event Upsets (SEU) where a cosmic ray or solar energetic particle flips a bit of data from 0 to 1 or vice-versa. Error Detection and Correction (EDAC) was used by the recorders to deal with the SEU problem. This involved encoding each word of 32 data bits into 39, with the other 7 bits for a Hamming code. This format allowed detection and correction of single-bit errors, and detection without correction of double-bit errors. The ASICs marched through the whole memory about once every 9min, checking each 39-bit word for consistency and correcting single-bit errors.

AACS, CDS, spacecraft telemetry and instrument memory loads were stored in separate files known as partitions. All data recorded to and played back from the SSR was handled by CDS. There were three different SSR functional modes: Read-Write to End, Circular FIFO, and Ring Buffer. There was also a record pointer and a playback pointer to indicate the memory addresses at which the SSR could write or read.

In Read-Write to End, there was a logical beginning and end to the SSR. Recording began at this logical beginning and continued until either the SSR was reset (the record and playback pointers were returned to the logical beginning) or until the record pointer reached the end. If the record pointer did not reach the end, recording was halted until the SSR was reset. In Circular FIFO there was no logical end to the SSR, and the data was continuously recorded until the record pointer reached the playback pointer. The Ring Buffer mode was similar to the Circular FIFO except that recording did not stop when the record pointer reached the playback pointer.

The **Antenna Subsystem (ANT)** provided a directional High-Gain Antenna (HGA) with X-, Ka-, S- and Ku-band for transmitting and receiving on all four bands. Because of its narrow half-power beam width of 0.14° for Ka-band, the dish had to be accurately pointed. There were in fact five Ku-band feed horns, to generate the five overlapping beams used in rapid succession to generate a usefully wide illuminated swath for RADAR imaging. The HGA and LGA1 (which was located on the HGA feed structure) were provided by the Italian Space Agency. The second Low-Gain Antenna (LGA2) was positioned below the probe and pointed in the –X direction.

The giant 4m wide HGA structure used ribs and rings to stiffen the parabolic reflector. Initially the dish was to have been fabricated

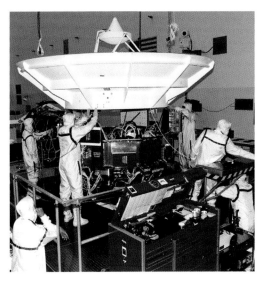

using a sandwich of carbon fibre skins and a Kevlar honeycomb. However, the strong thermal gradients expected while the spacecraft was in the inner solar system, when the HGA would be Sun-pointed in order to provide shade for the body of the spacecraft, drove the designers to use an aluminium honeycomb. The intense sunlight just within the orbit of Venus (3,500W/ m^2 at 0.625 AU from the Sun, more than double that at Earth) meant that the materials and adhesives had to tolerate temperatures of 140°C and intense ultraviolet light in the inner solar system, as well as temperatures below –200°C at Saturn. Several types of white paint were tested for their high temperature behaviour and ability to sustain the ultraviolet flux near the Sun without yellowing or otherwise degrading:

LEFT 'A good set of tools is essential.' A NASA toolbox is visible in the foreground as technicians affix the brilliant white HGA to the Upper Equipment Module. The cone at the apex of the antenna assembly housed a Low-Gain Antenna and the S-band antenna feed. Its base was the sub-reflector (like a secondary mirror in a telescope) for the X-, Ka-, and Ku-band beams. *(NASA/KSC)*

BELOW Schematic of the antenna system on Cassini. The High-Gain Antenna services the five-beam Ku-band radar experiment, as well as the S-band radio science and probe relay, and the Ka-band radio science. The High-Gain and two Low-Gain Antennas provide X-band telemetry and control, tracking and radio science. *(NASA/JPL)*

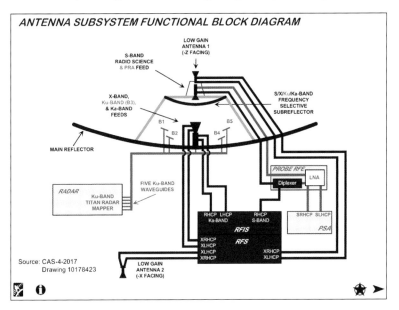

ANTENNA SUBSYSTEM FUNCTIONAL BLOCK DIAGRAM

PCBZ paint was chosen for the main structure with HINCOM for the supports and the subreflector.

One or other of the two LGAs was able to receive/transmit X-band from/to Earth when the spacecraft was Sun-pointed. The LGAs also provided an emergency uplink/downlink capability while Cassini was at Saturn (and were used for radio tracking on some of the later flybys). The HGA downlink gain at X-band was 47dBi and the LGA1 peak downlink gain was 8.9dBi. The X-band TWTA power was 20W.

Orbiter instruments

Twelve scientific instruments were carried by the orbiter. For seven of these, a principal investigator and coinvestigators were responsible for the instrument as well as the scientific investigation, with the other five being designated as 'facility instruments' that were built at JPL and were to serve a Cassini science team.

Instruments for remote sensing

Six of the orbiter instruments were to measure properties of targets at a distance: the Imaging Science Subsystem, Visible and Infrared Mapping Spectrometer, Composite Infrared Spectrometer, Ultraviolet Imaging Spectrograph, Cassini RADAR, and Radio Science. The first four of these were mounted and coaligned on the remote sensing pallet that was aimed in the +X direction of the spacecraft; the radio and RADAR investigations used the High-Gain Antenna (HGA) in the –Z direction.

The **Imaging Science Subsystem (ISS)** consisted of a Wide-Angle Camera, a Narrow-Angle Camera, and related electronics. Each camera had optics, a filter wheel, a shutter and detector head, plus electronics. In addition to acquiring scientific data, they were used for optical navigation and refining knowledge of Cassini's trajectory.

The Wide-Angle Camera (WAC) had refractive optics with a focal length of 200mm, a focal length-to-diameter ratio (f-number) of 3.5, and a 3.5° field of view.

The Narrow-Angle Camera (NAC) was a Ritchey-Chrétien reflecting telescope with a 2,000mm focal length at f/10.5, and a 0.35° field of view.

Filters were mounted in two rotatable wheels per camera. The WAC had 18 filters over the range 350–1,100nm (its glass lenses were spares from the Voyager camera, and they prevented transmission of ultraviolet shortward of 350nm). The NAC had 24 filters spanning the near-ultraviolet (200nm) to the near-infrared (1,100nm).

The sensing element of each camera was a 1,024x1,024 element CCD (advanced for its time!) that had 12μm pixels coated with phosphor to provide ultraviolet response and cooled to 180K by a radiator to reduce dark current. The dynamic range of each camera was about 4,000:1, which was equivalent to 12 bits. Two-blade focal plane shutters controlled the exposures. Fixed or automatic

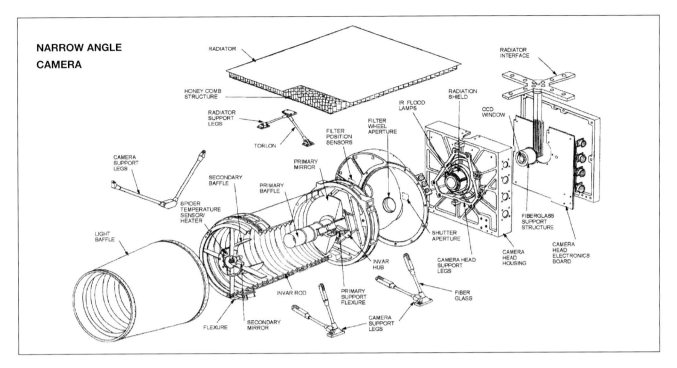

NARROW ANGLE CAMERA

RADIATOR

HONEY COMB STRUCTURE

RADIATOR SUPPORT LEGS

TORLON

CAMERA SUPPORT LEGS

SECONDARY BAFFLE

SPIDER TEMPERATURE SENSOR/ HEATER

LIGHT BAFFLE

FLEXURE

SECONDARY MIRROR

PRIMARY BAFFLE

PRIMARY MIRROR

FILTER POSITION SENSORS

FILTER WHEEL APERTURE

IR FLOOD LAMPS

RADIATION SHIELD

RADIATOR INTERFACE

CCD WINDOW

INVAR HUB

INVAR ROD

PRIMARY SUPPORT FLEXURE

SHUTTER APERTURE

CAMERA HEAD SUPPORT LEGS

FIBER GLASS

CAMERA SUPPORT LEGS

CAMERA HEAD HOUSING

FIBERGLASS SUPPORT STRUCTURE

CAMERA HEAD ELECTRONICS BOARD

ABOVE The NAC of the ISS was essentially a small version of the Hubble Space Telescope with a focal length of 2m and an f/10.5 Ritchey-Chrétien design. *(NASA/JPL)*

RIGHT The NAC of the Cassini ISS during bench tests. *(R. West/JPL)*

BELOW The WAC for ISS coupled a spare Voyager refractive optics (compromising its ultraviolet sensitivity but saving cost) to a CCD detector assembly that was identical to that of the NAC and a filter wheel assembly. *(NASA/JPL)*

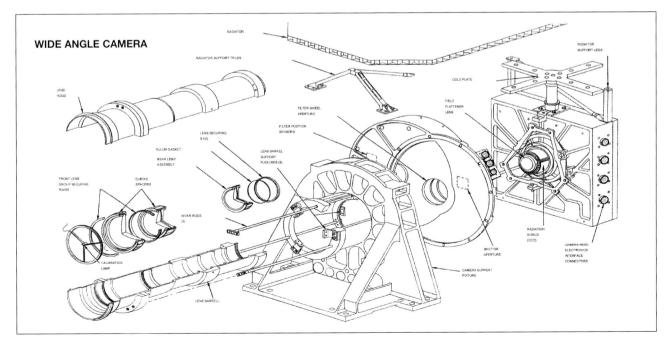

WIDE ANGLE CAMERA

RADIATOR

RADIATOR SUPPORT TRUSS

LENS HOOD

LENS SECURING RING

RULON GASKET

REAR LENS ASSEMBLY

FRONT LENS GROUP SECURING RINGS

DUROID SPACERS

INVAR RODS (3)

CALIBRATION LAMP

LENS BARREL

FILTER WHEEL APERTURE

FILTER POSITION SENSORS

LENS BARREL SUPPORT FLEXURES (6)

COLD PLATE

FIELD FLATTENER LENS

RADIATOR SUPPORT LEGS

RADIATION SHIELD (CCD)

SHUTTER APERTURE

CAMERA SUPPORT FIXTURE

CAMERA HEAD ELECTRONICS INTERFACE CONNECTORS

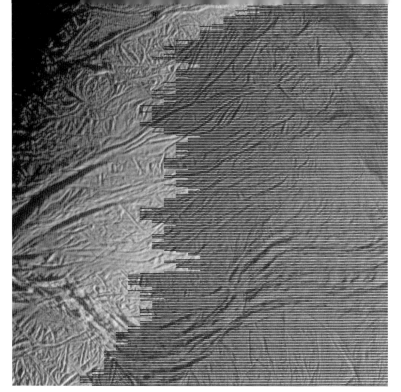

ABOVE A lossy run-length-encoding compression mode for Cassini's imager produced an odd appearance in this image of fissures on Enceladus. The blocky boundary was determined by how much information was in each row of the image (working from the left). The allocation of bits for each row was used up rapidly if there were lots of bright/dark transitions (e.g. at bottom) but shadowed or smooth areas got further to the right before the bits ran out. This compression approach was very efficient when large regions of the image were black. In entropy-rich scenes (as this one) where the bits got used up too rapidly, reconstruction of the right half of the image would be impossible. To avoid this, only every other row was compressed, to ensure the whole scene was still visible. *(SSI/JPL/NASA)*

exposure control modes could be used, and images could be binned 2x2. The data could be reduced to 8 bits/pixel by a square-root compression lookup table or by reading the 8 least-significant bits. A lossless data compressor with run-length encoding provided an average compression of 2:1. Higher compression ratios could be applied, but with loss of information.

The *Visual and Infrared Mapping Spectrometer (VIMS)* was to furnish information about the surface and atmospheric composition of Saturn and its satellites. It would provide images in which every pixel contained a high-resolution spectrum of the corresponding spot on the ground. VIMS mapped the areas viewed with lower spatial resolution than ISS, but at 352 contiguous wavelengths between 0.35–5.1μm.

The 0.85–5.1μm infrared channel employed a 230mm aperture f/3.5 Ritchey-Chrétien telescope whose secondary mirror scanned in two axes to build a 64x64 pixel image covering 1.9x1.9°. The light that was gathered by the telescope was passed through a shutter, spectrometer slit, and collimator to a diffraction grating that spread the light into a spectrum which was imaged, via reflective optics, by a linear array of 256 indium antimonide

ABOVE The VIMS instrument had two telescopes and diffractive spectrometers. At the top here is a (little used) visible channel provided by Italy. At the bottom is the JPL-developed near-infrared channel which had a larger telescope and a radiative cooler. *(NASA/JPL)*

LEFT The assembled VIMS instrument. The large gold-covered plate located on the right is the radiator for the detector cooler. *(NASA/JPL)*

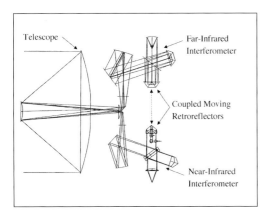

Telescope

Far-Infrared
Interferometer

Coupled Moving
Retroreflectors

Near-Infrared
Interferometer

ABOVE The CIRS spectrometer had an interferometer to perform wavelength selection. A stage oscillated back and forth to change the optical path length on the scale of the wavelength being sensed in the range 5–500µm. *(NASA/GSFC)*

ABOVE The focal plane assembly of the CIRS instrument. Light entered through the cone (bottom) to reach the detectors on the small gold-plated block (centre), which was held in place on small cylindrical titanium legs. The flexible curved copper thermal strap (top) connected the detectors to the radiator to keep them cool, while electrical connections were made with copper traces on a thin Kapton tape to minimise heat leaks. *(P. Irwin)*

photodiodes. The higher order spectral contributions were blocked by filters on a sapphire substrate in front of the detectors. The detectors were cooled by a radiator to 56K to minimise noise.

The visible channel (supplied by Italy) observed in the range 0.35–1.05µm using a holographic grating spectrometer that dispersed the light onto a 2D silicon CCD detector, one dimension being in space, the other being 96 wavebands. The primary mirror could be scanned around the other spatial dimension in order to build up spectral images with the same 0.5x0.5mrad pixels as the infrared channel.

The **Composite Infrared Spectrometer (CIRS)** measured thermal infrared radiation from 7–1,000µm. The assembly included a 51cm f/6 Cassegrain telescope with a 7.6cm hyperboloidal secondary mirror, three interferometers, and an 80K cooler.

The far-infrared interferometer employed a polarising beam splitter using wire grids and two thermopile detectors. The mid-infrared (7–17µm) channel used a Michelson

RIGHT The CIRS instrument mounted a telescope assembly (visible at the top) to a rigid interferometer optical bench and detector assembly (middle) and passive radiative cooler plate (left). *(D. Jennings/NASA/GSFC)*

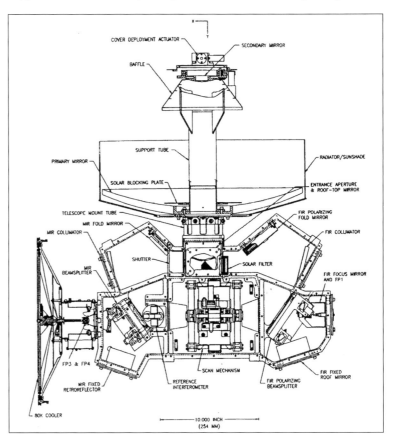

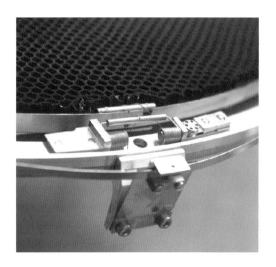

RIGHT A close-up of the CIRS cooler plate, showing part of the mechanism that released the cover. The surface of the cooler was coated with a black honeycomb material to maximise its ability to radiate heat to space and maintain the detectors at a temperature of ~60K. *(S. Calcutt)*

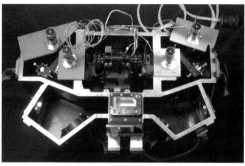

ABOVE The stiff housing of the interferometer for CIRS was milled from a single piece of aluminium and optically dark coatings applied internally to minimise reflections. *(S. Calcutt)*

LEFT The CIRS analogue-to-digital converter board (left) shows the tidy ordered layout of the surface-mount integrated circuit. Part of the analogue circuitry (right) associated with the FP1 detector shows the blobs of adhesive on several wires and components, as well as the white strap around the electrolytic capacitor – these measures were to enable the electronics to withstand the severe vibrations of launch. *(S. Calcutt)*

BELOW Build-up of the CIRS telescope assembly – a lightweight beryllium mirror support, with a polished surface which was then gold-coated to maximise its infrared reflectivity, became part of a measurement system integrated with sensors and electrical wiring. *(S. Calcutt)*

BELOW Testing the CIRS detectors and cooler in a chamber at Oxford University. Liquid nitrogen was used to chill the chamber walls to the cold of space. *(P. Irwin)*

ABOVE **The CIRS telescope, integrated with its cooler for ground tests.** *(P. Irwin)*

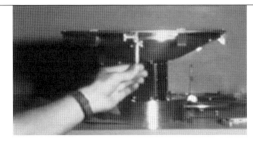

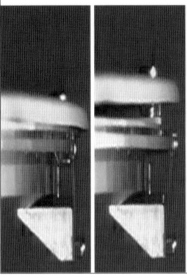

interferometer and a germanium lens to focus the radiation on two focal planes, each with linear arrays of mercury cadmium telluride detectors operating in photovoltaic and photoconductive modes. A third interferometer using a 785nm laser diode generated a servo signal for the motor-driven mechanism that moved reflectors back and forth to change the path length of the interferometers and thereby scan in wavelength.

The mid-infrared detector arrays were mounted on an 80K cooler. Other portions of the optics assembly were cooled to 170K by a separate radiator. To reduce heat leakage into the cold optics assembly, special wires with low thermal conductance were used for its electrical leads.

ABOVE **A spring-loaded cover protected the CIRS radiator from accidental exposure to the Sun while Cassini was in the inner solar system, and was released during the cruise to Jupiter. A ground test, actuated by hand instead of the paraffin wax actuator which would be used in-flight, used a high-speed camera to verify that the cover would be ejected at the required speed of 1m/s.** *(P. Irwin)*

LEFT **The delivery and integration of a science instrument requires detailed planning of tasks and items ranging from the prosaic but crucial issues of shipping containers and cranes, to the engineering of handling fixtures (like the C-shaped fixture that allowed precision rotation of the remote sensing pallet during integration at KSC) to the surgical matters of nitrogen purges, and finally, the removal of pre-flight covers.** *(S. Calcutt)*

Like other instruments developed for Cassini, UVIS was a composite of several sub-instruments, each remarkably sensitive and capable, with components coming from many countries.
(U. of Colorado)

BELOW Like the VIMS instrument, UVIS used diffraction gratings to spread the different wavelengths of light out across the detector.
(U. of Colorado)

BELOW RIGHT The assembled structure of the UVIS instrument. Note the milled cut-outs to save weight without compromising structural stiffness.
(U. of Colorado)

The **Ultraviolet Imaging Spectrograph (UVIS)** instrument measured ultraviolet emissions at brightness levels of 0.001 Rayleigh to several thousand Rayleighs. (One Rayleigh is an emission rate of 10 billion photons/m²/s. The night sky of Earth has an intensity of about 250R, while auroras can reach values of 1,000kR.) It was a two-channel spectrograph with Far-Ultraviolet (FUV) and Extreme-Ultraviolet (EUV) channels, and had a Hydrogen Deuterium Absorption Cell (HDAC) and a High-Speed Photometer (HSP).

Each of the two spectrographic channels utilised a reflecting telescope, a concave grating spectrometer, and an imaging, pulse counting detector. The telescope primary was an off-axis parabolic section with a focal length of 100mm, a 22x30mm aperture, and a field of view of 3.67x0.34°. The aberration-corrected toroidal grating focused the spectrum onto an imaging Microchannel Plate (MCP) detector.

The Far-Ultraviolet channel had a wavelength range of 115–190nm. Selectable entrance slits provided a resolution of 2.4Å. The range of the extreme-ultraviolet channel was 55–115nm. This channel also had a port to permit sunlight from 20° off-axis to enter the telescope for occultation observations.

The High-Speed Photometer with a 0.34x0.34° field of view measured light (115–185nm, undispersed) from its own parabolic mirror with a Hamamatsu R1081 photomultiplier tube detector with a caesium iodide photocathode. This instrument could sample light levels rapidly (500 times/sec) to resolve fine structure in the rings or other occultation targets.

The Hydrogen Deuterium Absorption Cell was a photometer to measure concentrations of hydrogen and deuterium. Behind an objective lens, it had three resonance absorption cells and an electron multiplier detector. One cell was filled with hydrogen and another with deuterium. A tungsten filament in each of these cells dissociated a fraction of the molecular gas and allowed the hydrogen-to-deuterium spectrum near the resonance lines to be measured to high resolution by varying the filament temperature. An oxygen cell was intended to act as a broadband filter.

The **Cassini Radar (RADAR)** was designed primarily to investigate the surface of Titan during close flybys of that satellite. It operated at

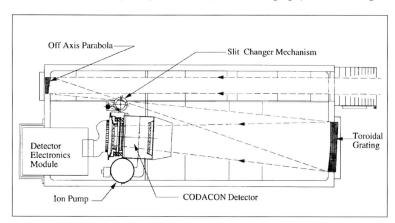

Ku-band (2.2cm wavelength, or 13.8GHz) and included a radio frequency electronics subsystem, a digital subsystem, and a power conditioner.

The radar used five microstrip array Ku-band feeds to illuminate Cassini's main 4m antenna. For long-range observations and altimetry only the central 0.3° boresight beam was used, whereas for Synthetic Aperture Radar (SAR) mapping close to Titan the radar was switched rapidly between the five beams, four of which (1.3x0.3°) were slightly offset to build an overlapped image swath typically 200–400km wide.

Range and Doppler processing of echoes from the chirped signals allowed the recovery of surface radar reflectivity maps with a resolution of down to 350m.

In its altimetry mode, the radar could measure the height profile of terrain beneath the spacecraft with a vertical precision of the order of 10–20m.

The peak power of the instrument was about 63W, and a capacitor bank buffered energy for individual pulses which had a much higher instantaneous power.

When used without transmitting, the receiver could be used as a radiometer to detect microwave emission from targets, measuring their brightness temperature to within a fraction of a Kelvin. (Cassini's Phase A 'model' payload had included an elaborate microwave spectrometer, but no such instrument was selected. Adding a somewhat improvised radiometry capability to the RADAR addressed at least a little of the science that was lost.)

The Orbiter **Radio Science (RS)** instruments were to provide data on the atmospheres and ionospheres of Saturn and Titan, on the rings, and on the gravity fields and orbits of the planet and its satellites. During the interplanetary cruise, the instrumentation were to be used to search for gravitational waves.

The RS investigation employed both the X-band communications link of the spacecraft RFS and the Ka-band and S-band capabilities of the Radio Frequency Instrument Subsystem (RFIS) – which included a transponder, a Travelling Wave Tube Amplifier (TWTA), and an ultra stable oscillator.

The RFIS contained an S-band transmitter and a suite of Ka-band equipment, namely a

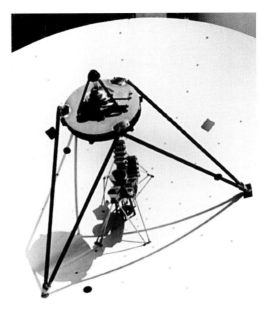

LEFT The feed structure for the High-Gain Antenna during a beam measurement test that was conducted outdoors (note the undressed cone at the apex, and the shadow of the feed). Two serrated rectangular Ku-band radar antenna feeds are visible slightly offset from the centre. *(ASI)*

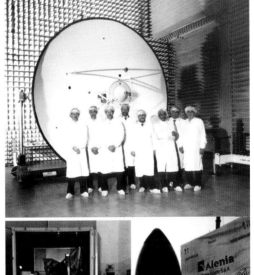

LEFT The formidable 4m diameter High-Gain Antenna was manufactured in Italy. The walls of the radio test chamber were covered in prismatic cones of radar-absorbing material to prevent reflections from disturbing test results. *(E. Flamini)*

LEFT The HGA was shipped by air from Italy to JPL. *(E. Flamini)*

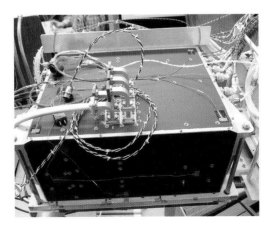

LEFT The Cassini RADAR instrument transmitted and received microwave radiation at a wavelength of 2.2cm. The centimetre-sized waveguides for the five beams can be seen at the top of the instrument electronics box. *(NASA/JPL-Caltech)*

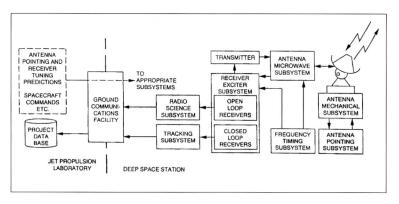

phase-lock loop translator that received and translated the uplink carrier by a factor of 14/15 for retransmission to Earth, an exciter that generated a downlink signal, and a Ka-band TWT amplifier. Transmission and reception were by the HGA. Downlinks with accurately known frequency could be transmitted using the X-band TWT amplifier, the Ka-band exciter and amplifier, and the S-band transmitter. Two-way coherent signals were achieved by X-band uplink to the transponder and by Ka-band uplink to the translator in conjunction with the X-band and Ka-band downlinks.

The system capabilities included one-way and two-way Doppler, differential one-way ranging, and two-way ranging. The links were with 34m and 70m antenna Earth stations of the Deep Space Network. The radio frequency output at Ka-band and S-band was 10W. Frequency stability of transmissions, set by the ultra stable oscillator, was better than 0.1 parts per trillion for integration times of 10–10,000sec. The two-way ranging provided an accuracy of 20–30nsec (6–9m).

Instruments for fields, particles and waves

The six orbiter instruments to measure properties of fields, particles, and plasma waves were the Dual Technique Magnetometer, Radio and Plasma Wave Science instrument, Cassini Plasma Spectrometer, Magnetospheric Imaging Instrument, Cosmic Dust Analyser, and Ion and Neutral Mass Spectrometer.

The **Dual Technique Magnetometer (MAG)** measured the magnetic field vector. It consisted of a three-axis Flux Gate Magnetometer (FGM) and a Vector/Scalar Helium Magnetometer (V/

SHM); the latter being a refurbished spare from the Ulysses mission.

The FGM was mounted near the midpoint of the magnetometer boom and the V/SHM at its tip. Using a combination of two magnetometers at different points helped in separating ambient fields from the field generated by the vehicle itself. The flux gate could operate over a wide range and measure rather rapidly (up to 30Hz), whereas the complementary helium sensor gave better low-field sensitivity and long-term calibration stability.

The FGM used three identical but orthogonal single-axis sensors with a permeable ring core wound with an excitation coil at 18kHz to drive the core into saturation. An external field caused the saturation cycle to become asymmetrical, which was sensed with a pickup coil.

The V/SHM used a low pressure helium lamp excited by a radio frequency signal (rather similar to the Huygens USO) but exploited the Zeeman splitting of the resonance line. The optical pumping of the cell was maximised when no field was present. By sweeping a known field applied using a set of Helmholz coils to null the ambient field, the latter could be recovered by sensing the increased optical absorption.

Away from the warmth of the spacecraft, the FGM incorporated an electrical heater and three RHUs in order to maintain its temperature at Saturn, while the VHM used electrical heating only. The relatively undemanding instrument operations were performed by an 80C86 processor whose power supplies and CPUs were redundant.

In order to minimise the effect of the spacecraft on the measurement – the specification was to limit the spacecraft contribution to the measured field to no more than 0.2nT, in contrast to the field near Earth's surface of 40,000nT! – not only was the instrument extended on a boom, but an inventory was made of the magnetic field of the different components.

Some 56 Cassini units were measured, and in some cases compensation magnets were installed (notably the propulsion system latch valves) and in other cases (like the reaction wheels) shields were added. In particular, the RTGs, despite having internal wiring designed to self-compensate for the fields made by their electrical current, contributed ~40% of the field at the sensor. A magnetic cleanliness program was instituted to measure the fields of each of the fuelled RTGs for the mission. This was done in a special concrete-walled facility in order to minimise personnel exposure to radiation. The mounting angles of the RTGs were then optimised, taking into account the lower currents that were expected in 2004 compared to the measurements in 1995–97, such that the residual fields of the three RTGs cancelled each other out. In the end, the specification was met (0.114nT) with margin.

The Spacecraft Alignment System (SCAS) included electrical coils that could be energised periodically in order to add a field (~20nT at the FGM) whose orientation relative to the vehicle was known, to verify the orientation of the sensors in flight (e.g. if the boom were to become deformed). Tests using these coils in 1999 and 2000 after the boom deployment, showed that the FGM was within 0.3° and the VHM within 1° of the ideal. Occasionally the vehicle was rolled for further calibration, because rotating about the Z axis separated out the contributions of ambient and spacecraft fields.

The **Radio and Plasma Wave Science (RPWS)** instrument measured varying electric and magnetic fields in the vicinity of the spacecraft, as well as electron density and temperature.

The electric fields were sensed by three long antennas, using collapsible 10m-long metal tubes that were extended by a motor after launch. The three antennas were arranged as a dipole and a monopole.

RPWS could perform direction-finding on radio sources. It could also detect the charge pulses generated by dust impacts on the antennas. The magnetic search coil assembly included three orthogonal coils about 25mm in diameter with 10,000 turns of 0.07mm wire around a 25cm-long Mu-metal core to detect fields from 1Hz to 20kHz.

Additionally, there was a Langmuir probe consisting of a 5cm titanium sphere mounted on a short (0.8m) hinged boom to measure electron density and temperature. The sphere was coated with titanium nitride to give a durable finish with a constant work function to improve its calibration stability, recording electron densities of 5–10,000 electrons/cm^3.

Electron densities could also be measured by the 'sounder' mode of the RPWS in which it would transmit square wave pulses in order to stimulate plasma resonances.

Signals from these antennas and electrodes – mounted just behind the HGA – were analysed by a suite of receivers with a dynamic range exceeding 90dB that could measure magnetic fields and electric fields in the ranges 1Hz to 12.6kHz and 1Hz to 16MHz respectively.

Various receiver circuits could 'listen' to the antennas, depending on whether long surveys (with limited temporal and spectral

ABOVE The helium vector magnetometer was mounted just above the three antennas of the RPWS experiment (with red covers). A tripod supported a box array of Helmholtz coils that surrounded the magnetometer sensor. By feeding varying currents through the parallel coils, the sensor was stimulated with known magnetic fields. (NASA/JPL)

LEFT An engineer checks the stowed MAG boom assembly, with the inboard (flux gate) magnetometer visible inside a rigid white truss between the two deployable segments that were covered in gold thermal blankets. (NASA/JPL-Caltech)

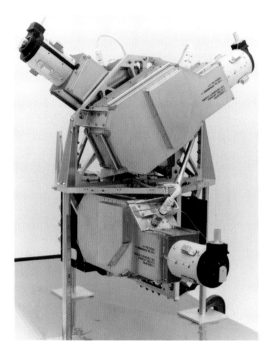

RIGHT The three antennas of the RPWS provide a dipole and monopole to sense electric fields. They were extended after launch in near-orthogonal directions and permitted the source direction and polarisation of electromagnetic waves to be measured. *(NASA/JPL)*

ABOVE A deployment test of an RPWS antenna extending from its spool canister. A pair of prestressed beryllium-copper tapes were stored flat on spools and unwound by a motor through a shaped channel to enable them to curl together to create a relatively stiff tube. The antennas were silver-plated on the outside, black painted inside, and with about 12% perforations to let sunlight in to heat the shaded side and thereby minimise thermally induced bending. *(NASA/JPL)*

resolution) were being performed, or high-bandwidth 'waveform' captures, such as dust impact signals. The instrument used three 80C85 processors to sequence its tasks and to undertake Fourier analysis and data compression on the signals. At its highest rate, RPWS could generate 365kbps of data!

The *Cassini Plasma Spectrometer (CAPS)*

measured the composition, density, flow, and temperature of ions and electrons in Saturn's magnetosphere. It comprised three

units: an Ion Mass Spectrometer (IMS), an Ion Beam Spectrometer (IBS), and an Electron Spectrometer (ELS). A motor-driven actuator rotated the sensor package to provide 208° scanning in azimuth about the symmetry axis of the orbiter.

The ELS used a 'top hat' curved-plate electrostatic analyser and Microchannel Plate (MCP) detectors for electron energy measurements. The ELS energy range was 0.7–30,000 eV with a resolution of 0.17. The sensor's field of view was 1.5x160° and the angular resolution was 5x20°.

The IBS used a hemispherical curved-plate electrostatic analyser and channel electron multiplier detectors to determine energy-to-charge ratios. The energy range of the IBS was 1eV to 50keV with a resolution of 0.015. The field of view was 1.5x160° and the angular resolution was 1.5x1.5°.

The IMS provided data on both energy-to-charge and mass-to-charge ratios. A curved plate electrostatic analyser provided energy-to-charge separation. The ions were

LEFT An RPWS test on a table. The tube could not support its own weight in terrestrial gravity, so the test deployed it horizontally on a long table with rollers at intervals of about 1ft to prevent the tube from buckling. *(NASA/JPL)*

then accelerated electrostatically to strike and penetrate a set of thin carbon foils. This gave rise to secondary electrons, and broke up some of the molecular ions. The secondary electrons struck an MCP and signalled the start of an ion's time of flight. The ions travelled in a chamber in which the electric field increased linearly along the analyser length. Positive ions having less than 15keV of kinetic energy were deflected back to the entrance end of the analyser where they struck an MCP, stopping the time-of-flight timer. Positive ions with higher energy, negative ions, and neutrals struck another MCP at the opposite end. The fragments from the break-up of molecular ions supplied information about their composition. The mass range of the IMS was 1–60amu with a resolution of 0.013.

The CAPS electronics included three PACE 1750A processors. Amptek A111F amplifiers/discriminators detected the tiny charge pulses of particles. Many of the internal surfaces of the instrument were coated in a special Ebanol-C copper oxide in order to minimise the reflection of ultraviolet light and particles. And special attention was paid to making the multi-layer insulation blankets nearby the instrument electrically conductive (by stitching with stainless steel wire) to minimise any charge build-up which could distort the CAPS measurements.

The **Magnetospheric Imaging Instrument (MIMI)** provided images of the plasma surrounding Saturn and determined ion charge and composition. Like CAPS, it incorporated three sensors.

One sensor, the Low-Energy Magnetospheric Measurement System (LEMMS), had ion-implanted solid state detectors to provide directional and energy information on electrons in the range 15keV to 10.5MeV, protons in the range 15–130MeV, and other ions with 20keV to 10.5MeV per nucleon. The LEMMS head was double-ended, with oppositely directed 15° and 45° conical fields of view. LEMMS was mounted on a platform that permitted continuous rotation of the head through 360° on an axis that was perpendicular to the HGA axis and to the LEMMS telescope axis.

Another sensor, the Charge-Energy-Mass

Cassini MIMI
FM LEMMS Sensor
19 July 1996
0 1 2 3 4 5 6 7 8 9 10
centimeters

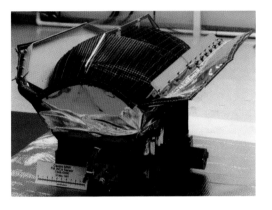

Spectrometer (CHEMS), measured charge and composition of ions in the range 10–265 keV using an electrostatic analyser, a time-of-flight mass spectrometer, and an MCP. The mass-to-charge range was 1–60amu/e (elements hydrogen to iron) and the molecular ion mass range was 2–120amu.

The third MIMI sensor, the Ion and Neutral Camera (INCA), was a time-of-flight camera with collimator slits, an entrance foil, and MCPs. It recorded ions and neutral molecules with energies of 10keV to about 8MeV per nucleon and provided remote images of the energetic neutral emission of the magnetosphere of Saturn. Although MIMI was listed as a fields and particles instrument, INCA might also be classified as a remote sensor because it produced images with particles instead of photons and had an angular resolution of about 2x2°.

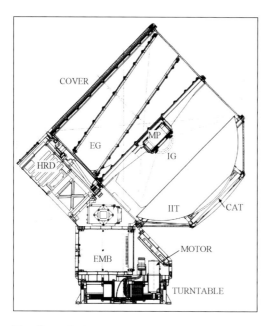

The **Cosmic Dust Analyser (CDA)** measured the flux, velocity, charge, mass, and composition of dust and ice particles in the mass range 10^{-16}–10^{-6}g, corresponding to particles ranging in diameter between about 1μm and 1nm.

The CDA combined two separate systems: a High-Rate Detector (HRD) and a Dust Analyser (DA).

The two HRDs were intended for measurements in Saturn's rings, and were able to count up to 10,000 impact/sec. They relied on charge pulses produced by impacts on an electrically polarised plastic film, polyvinylidene fluoride (PVDF). One detector had a 50cm² area with a 28μm thick film, and the other was only 10cm² with a thinner and more sensitive (6μm) film. To prevent high-temperature degradation of the detectors during the warm inner solar system cruise, the films were coated with Chemglaze Z-306 white paint. Absorbing mounts isolated the detectors from vibrational 'noise' in the spacecraft structure. Circuits counted the charge pulses above a set of four thresholds, each corresponding roughly to a given particle momentum.

The DA determined the electric charge carried by dust particles, their flight direction and impact speed, mass, and chemical composition, at rates of up to 1 particle/sec and for speeds in the range 1–100km/s. Two pickup electrode grids at its entrance registered particle charge. Charge-sensitive amplifiers and a logarithmic amplifier permitted recording over a range of 10^{-16}–10^{-12} coulombs. Most of the incoming particles then struck an impact ionisation target held at a potential of 0V. An ion collector grid at 350V accelerated the positive ions of the impact plasma, with many of them passing through the grid to reach an electron multiplier detector. The timing of the signals produced by these three elements indicated the impact velocity, and the pulse heights indicated the dust particle mass.

Dust particles striking a separate chemical analyser target in the DA also produced impact ionisation. This target was held at +1,000V and had a grounded grid in front of it to accelerate positive ions. Ions reaching the grid signalled the start time for the time-of-flight mass spectrometer. Other ions passed through into the spectrometer and eventually reached the ion collector and electron multiplier. Their time of flight was an inverse function of the ion mass. The ion masses (atomic weights) were measured with a resolution of about 50, and their distribution gave the chemical composition of the dust particle. The chemical analyser target was made of rhodium (a rather exotic metal, unlikely to be present in dust particles and hence easy to separate from the composition of the dust particles).

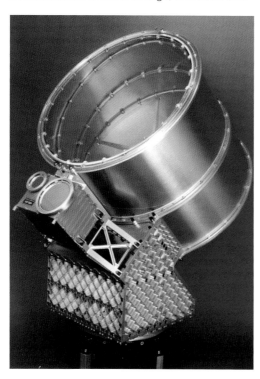

An articulation mechanism permitted the sensors to be rotated through 270°, typically at 10°/min using a Phyton ZSS32 stepper motor with plastic bearings, a 1,000:1 reduction gear, and Mu-metal shielding to minimise stray magnetic fields.

The DA data processing was performed by an MA31750 processor running at 6MHz (coded, like most MIL-STD-1750A processors, in the Ada language) that could perform a lossless Rice data compression (typically 3:1) or lossy wavelet compression.

The **Ion and Neutral Mass Spectrometer (INMS)** was to determine the chemical, elemental, and isotopic composition of the gaseous and volatile components of the neutral particles and the low-energy ions near the rings and in Titan's upper atmosphere.

It included two ion sources ('open' and 'closed'), plus a quadrupole mass analyser with an ion detector. A 90° quadrupole deflector selected ions from one of the two sources and fed them into the mass analyser, which used the same quadrupole electrodes as the Huygens GCMS.

The ions, being electrically charged, could be analysed directly as they entered, but the neutral molecules had first to be ionised using an electron gun. The open source (used for species that might react with the instrument surfaces) accepted neutrals within 8° of its boresight. The closed source, with a hemispherical field of view, enhanced the density of particles by up to 45 times by pointing an aperture forward so that the spacecraft velocity rammed particles into a cavity. The INMS could read masses up to 99 Dalton (i.e. 99 times the mass of a hydrogen nucleus, which is a single proton of 1amu).

BELOW The mass selection in the INMS instrument used four 152mm hyperbolically contoured electrode rods which were driven with radio frequency signals. The same rod design was used in the Huygens GCMS. *(NASA/GSFC)*

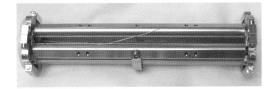

ABOVE The sampling head of the INMS accepted ions or neutrals from a range of angles and directed them into one of two inlets in the hemispherical inlet assembly, then the instrument ionised the neutrals and sorted the ions with a quadrupole mass analyser for counting. This view of the base of the instrument displays one of the electronics boards. *(NASA/GSFC)*

BELOW The two inlet apertures of the INMS – the spherical 'closed source' antechamber and electron gun (upper) and the 'open source' (below). On the right is the ion beam switching lens which selects which source the measured ions come from. *(NASA/GSFC)*

Chapter Four

Assembling Cassini-Huygens

Efforts across two continents to make Cassini, Huygens and their instruments ready for flight would culminate at Cape Canaveral in Florida in 1997, where they would be mounted – more than once – on their giant Titan launcher. ESA and many US government departments would be involved, and the project drew the interest of hundreds of thousands of people.

OPPOSITE After Cassini and Huygens were joined securely together at the launch site, their mechanical and electrical links would hold as designed for seven long years in space until explosive bolts and powerful springs effected separation. *(ESA)*

Final tests and assembly

The final construction of Cassini, known as ATLO (Assembly, Test and Launch Operations) began in November 1995 with the structural components and cable harness. By April 1996 the orbiter engineering subsystems and some science instruments were assembled, and by early August the rest of the science payload was installed for a long campaign of testing. Some of these included the Huygens EM or STPM. During tests at instrument level, problems such as cracked structures and leaky valves led to the replacement of some instruments by their flight spares, highlighting the importance of these items.

Tests, common to most spacecraft, included electromagnetic compatibility tests (i.e. measuring the radio emissions from equipment and testing their susceptibility to interference), thermal-vacuum testing, and vibration and acoustic testing. These latter tests were carried out in November 1996 by subjecting the Cassini spacecraft to the simulated thundering noise and rattling vibration of a Titan launch. The thermal-vacuum tests in January and February 1997 used a 25ft diameter space simulation chamber at JPL. After pumping down the chamber, the spacecraft was maintained at high temperatures to drive off (outgas) any volatile surface contaminants as a prelude to four days of testing using different levels of intensity from xenon lamps which simulated the varying solar heating that Cassini would experience, in order to verify that it could control its temperatures.

Cassini's integration and testing was performed at JPL in California. Although the prime contractor for Huygens was Aerospatiale of Cannes, France, the final integration of the Huygens probe was undertaken at Daimler Benz Aerospace Dornier Satellitensysteme in Ottobrun, near Munich, Germany. Much of the environmental testing was done at nearby Industrie Anlagen Betriebes-Geselleschaft (mercifully abbreviated to IABG).

In mid-1997, with launch imminent, the focus of Cassini activities progressively transferred from JPL and Europe to the Kennedy Space

BELOW A technician secures one of the hundreds of electrical cables on Huygens. The PDD and parachute box are already installed and the parachute lines secured with velcro tunnels. The aft cone has not yet been attached, and red preflight covers are seen on the DISR instrument and a HASI boom. *(P. Couzin)*

BELOW RIGHT The structural model of the Cassini spacecraft ready for acoustic tests at JPL. Note the microphones on booms. *(JPL-Caltech)*

ABOVE The Huygens STPM on a vibration table for shake-testing at IABG in Germany. The red box at the upper left trails cables to dozens of accelerometers to measure the structural response of the vehicle to a vibration spectrum simulating a Titan launch. (S. Lingard)

RIGHT Cassini in a thermal-vacuum chamber for tests at JPL. Brilliant xenon lamps at the top of the chamber simulate intense sunlight. The MAG boom is installed separately on a trestle in the foreground. (NASA/JPL)

BELOW Cassini in a silvery plastic shroud, on a truck at the gates of JPL. (NASA/JPL)

LEFT Cassini arriving at KSC in its air-conditioned container aboard a USAF C-17 jet transport. (NASA/KSC)

Center (KSC) in Florida and – because Titan IV launches were conducted by the 45th Space Wing of the Air Force – the adjacent Cape Canaveral Air Force Station (CCAFS).

First to arrive was Cassini's propulsion module subsystem in March 1997 onboard an Air Force C-17 jet transport. In the Spacecraft Assembly and Encapsulation Facility (SAEF-2) the pyrotechnic valves were installed, the helium tanks were pressurised, and so on.

The Space Shuttle program was busy at this time. On 4 April 1997 Columbia was launched for what was intended to be a 16-day microgravity research flight, but a faulty fuel cell caused the flight to be cut short and it landed back at KSC four days later. Around this same time, a Lufthansa cargo jet arrived at CCAFS carrying the Flight Model of the Huygens probe. The probe, in a pressure-tight shipping container, was moved to the cavernous Payload Hazardous Servicing Facility (PHSF) for unpacking and preparation for space flight.

Cassini itself arrived in Florida on 21 April 1997 after being trucked 90 miles from JPL to Edwards Air Force Base and then flown to KSC aboard an Air Force C-17 jet. The RTGs arrived discreetly, having been transported from the Department of Energy's Mound laboratory in Idaho by land for safety and security. Even the date on which Cassini's RTG's arrived at KSC was not publicly disclosed.

Planetary exploration has been compared

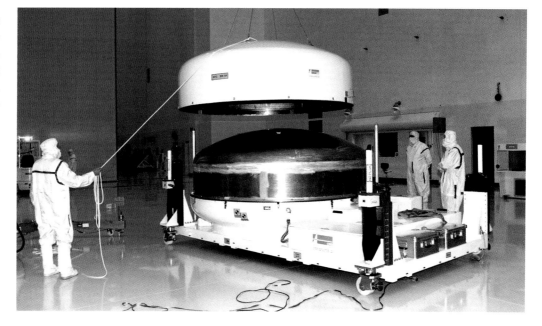

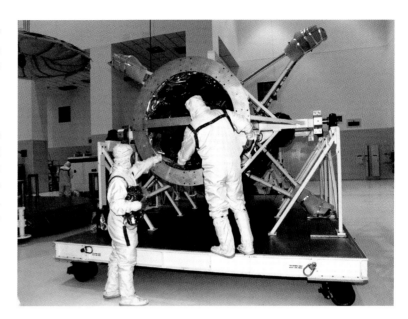

to the ancient construction of the Pyramids and the building of medieval cathedrals, in that it is an expensive but ultimately worthwhile human endeavour that harnesses the efforts of thousands in a transcendent project of lasting value to humanity. And in fact, Cassini in its shipping container made part of its journey under human power much like a stone block destined for the Pyramids, when it was manhandled into the PHSF.

In the PHSF, the lower section of the spacecraft was mounted on the launch vehicle adapter. The upper equipment section and HGA were then assembled. In May, the PMS was transferred to the PHSF and the spacecraft finally took shape – the two-storey tall spacecraft now needing a gantry to support personnel around it. The remote sensing pallet, with its various instruments, was lifted up and guided into place.

The RTGs were removed from their special transport containers and attached to Cassini for mechanical and electrical checks. The RHUs were installed on Huygens. Even though the actual radiation dose was small, the sensitive detectors on DISR began to show tiny changes in sensitivity as neutrons emitted by the generators penetrated their delicate semiconductor lattices, knocking an atom out of position here or there. After these checks,

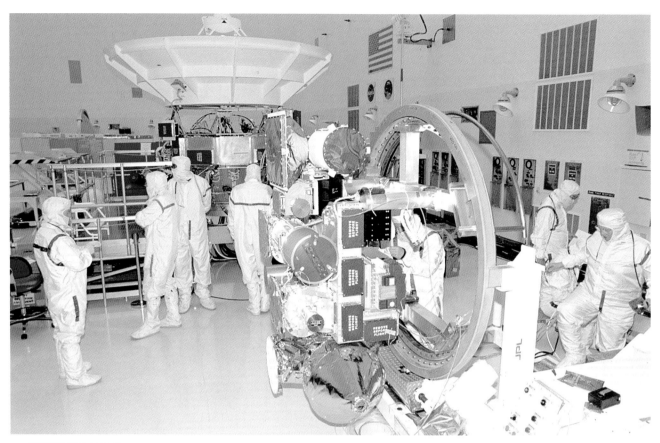

ABOVE The remote sensing pallet, installed on a rotating ground handling fixture, is being readied for attachment to the Cassini upper section. *(NASA/KSC)*

RIGHT The core stage and SRMU boosters of the Titan IVB launch vehicle for Cassini is pulled from the SMARF by a USAF locomotive, to begin its journey to Launch Complex 40 at Cape Canaveral Air Force Station. *(NASA/KSC)*

FAR RIGHT Cassini's launch vehicle, looking squat without its upper stage, arrives at the pad. *(NASA/KSC)*

FAR LEFT The Centaur upper stage in a protective cover is hoisted to the top of the gantry to be mated to the Titan core stage. *(NASA/KSC)*

LEFT The aft cone is lowered onto the Huygens probe in the payload handling facility at KSC. *(NASA/KSC)*

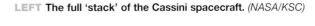

LEFT The full 'stack' of the Cassini spacecraft. *(NASA/KSC)*

BELOW Cassini's propulsion system, with black control unit and thruster outriggers visible, is mated to the lower equipment section. *(NASA/KSC)*

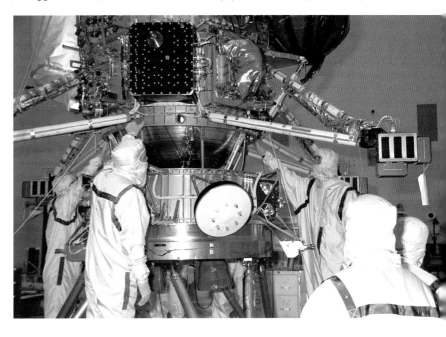

ABOVE The assembly of the Huygens probe is nearly complete, just awaiting installation of the back cover. A hose feeds cold air to prevent the probe from overheating during tests after the RHUs had been installed. Note the stack of books on the table: logbooks in which to record every connector mate and demate, and checklists to help prevent mistakes during assembly and test operations. Cassini is visible in the background. *(NASA/KSC)*

BELOW Removing an RTG from its transport cask to undertake mechanical and electrical checks. Notice the glass walls used to minimise radiation exposure of personnel. The Cassini spacecraft is visible in the background. The white suits are to protect Cassini from skin flakes and other biological debris, not to protect the personnel. *(NASA/KSC)*

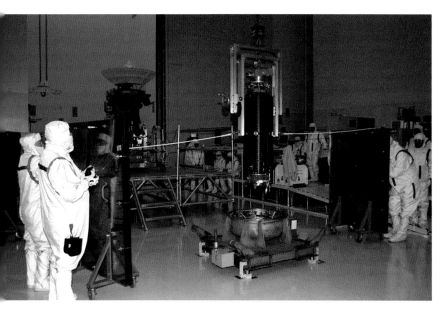

the RTGs were removed again because their emissions would complicate the assembly and test operations by requiring special precautions and monitoring (like dosimeter badges on personnel). The RTGs would be re-installed when Cassini was on the launch pad.

Thermal blankets were then attached to the naked aluminium structure of Cassini. All in all, some 250 pieces of multi-layer insulation were affixed. These were usually gold in colour, but some of them had black surfaces. The Huygens probe, covered in its own MLI blankets, was then hoisted into position on the Spin-Eject Device and the electrical connections mated. A final series of electrical tests were made.

The integrated Cassini-Huygens, on its launch vehicle adapter, was then covered with a protective bag and mounted on a truck with an aluminium transport cover. In the early hours of 29 August, complete with a security escort, the spacecraft was taken from the PHSF to Launch Complex 40. The aluminium cover was removed, and the spacecraft in its bag was hoisted up and mated to the Centaur by 08:15hrs.

In order to keep Huygens cool now that it had continuous heating by the RHUs inside its thermal covers and was immersed in the warm terrestrial environment, and in particular to prevent heat from degrading its batteries, a special air conditioning hose was to pipe in chilled air. Unfortunately, up on the top of the launch vehicle, the air conditioning unit was mistakenly set to supply air ten times faster than had been intended. An inspection using a borescope showed that this blast of air had shredded some of the foam insulation blocks inside the probe. Might a stray particle of foam material cause a mechanism failure at some point during the mission? With the launch date looming, the difficult decision was made to return Cassini to the PHSF. Round-the-clock work began to unbolt the launch vehicle adapter from the Centaur and to disconnect the umbilical connectors. Once again wrapped in its protective bag, Cassini was gently lowered onto the truck and driven back into the PHSF on 7 September.

Exercising procedures that doubtless engineers hoped never to have to use again, the pyros were safed, the pre-flight covers put back on, and the probe demated from Cassini

RIGHT The RTG connectors are mated for electrical tests. The RTGs were then removed and placed back in storage until just before launch. *(NASA/KSC)*

and opened up. The interior was inspected and the damaged foam repaired (in fact many of the best launch-site photos of Huygens come from this exercise). Fortunately only a few square centimetres of insulation had been damaged and were easily cleaned and repaired. After 111hr of hectic but disciplined operations, Huygens was reinstalled on Cassini and the spacecraft returned to the launch pad. After some struggles to complete the reintegration process, including consulting the Lockheed Martin Titan launch vehicle team in Denver, Colorado, who advised applying lubricant to the threads of the attachment bolts, on 16 September Cassini was ready for final rehearsals of the launch procedures.

On 10 October, the delicate 11hr process

BELOW Huygens is lifted into place and attached to the Cassini orbiter. *(NASA/KSC)*

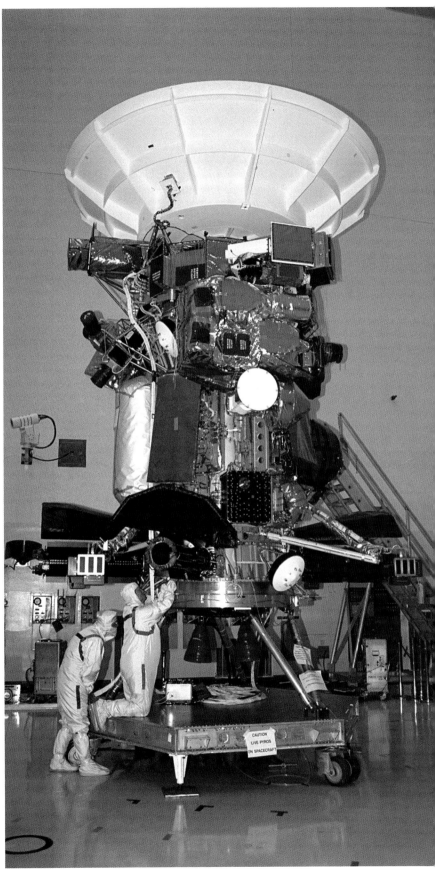

RIGHT The integrated Cassini-Huygens payload is lowered onto the launch adapter. *(NASA/KSC)*

FAR RIGHT 'It's a wrap.' A giant aluminised plastic bag is lowered over Cassini-Huygens to protect it during transport to the launch pad and while being hoisted to the top of the launch vehicle. *(NASA/KSC)*

RIGHT Cassini-Huygens on its launch adapter attached to the Centaur atop the Titan. A ground support 'diverter' box can be seen at the lower left, with a cold-air hose snaking up to Huygens. Unfortunately, the airflow through this hose was set too high by personnel on the pad. *(NASA/KSC)*

FAR RIGHT Having decided to open up Huygens, the first step was to winch Cassini/Huygens, once again wrapped in their protective bag, back to the ground. *(NASA/KSC)*

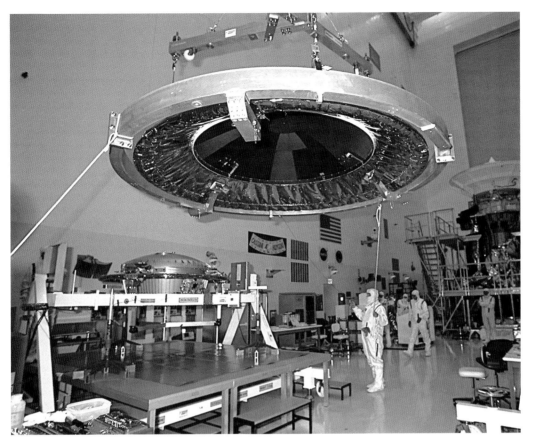

LEFT 'Disassembly is the reversal of assembly.' Dornier Satellitensysteme personnel remove the front shield from Huygens. *(NASA/KSC)*

BELOW Like forensic detectives, workers focus their attention with handheld lamps as they inspect the Huygens experiment platform for loose particles of foam. The Huygens batteries are nicely shown in this view. *(NASA/KSC)*

ABOVE Blocks of insulating foam, encapsulated in aluminised Kapton bags. Fortunately only a few blocks had small areas of damage and those were easily repaired. (NASA/KSC)

ABOVE RIGHT After reassembling Cassini/Huygens and re-mounting on the Titan, the aerodynamic fairing (black internal panels) are being readied for installation, to protect the payload during the ascent into space. (NASA/KSC)

BELOW 'Batteries not included.' Only 48hrs prior to the scheduled launch, the RTGs were installed through a special hatch in the Titan's payload fairing. (NASA/KSC)

RIGHT With everything finally ready, the mobile service tower was rolled away from the 56m-tall Titan. (NASA/KSC)

of transporting the RTGs to the pad and reinstalling them on Cassini and turning their power on (actually, it was a case of unshorting them because they never stop producing power) was performed. Then with the final memory uploads installed, the Cassini-Huygens mission was ready to go.

Power for Cassini

The availability of Radioisotope Thermoelectric Generators (RTG) that could provide non-stop power was essential to the practical operation of a well-instrumented and agile spacecraft at Saturn.

While satellites in Earth orbit can use solar arrays, and some modern space missions (e.g. Rosetta and Juno) have operated on solar power as far out as Jupiter, the inverse square law gives ~100 times less power per unit area at Saturn than at Earth, and four times less than at Jupiter. Not only would the massive solar arrays required be challenging to accommodate – even with 2017 array technology, let alone 1995 technology – but their tendency to 'flap' would make it impossible to turn the vehicle nimbly and retain precise pointing for science observations during brief target encounters. And in order to survive eclipses, a solar powered spacecraft would have needed a massive battery. Furthermore, it should be remembered that the 'waste' heat from the RTGs was important in preventing the propulsion system from freezing. In the absence of heat from the RTGs, the solar panels would have needed to be even bigger to provide the additional power required for electrical heaters. All these factors made solar power for a Saturn mission impractical.

Not every instrument on Cassini would run full-bore at one time – a number of operational modes were defined for Cassini, with different combinations of systems and instruments. But for any useful combination of Cassini's capabilities, two RTGs (as envisaged in some early spacecraft configurations) would have been insufficient. Three would be needed, each giving about 280W at launch. The three combined Cassini RTGs were required to yield 826W of power at the beginning of the mission and 596W by 16 years after launch (corresponding to 18 years after fuelling). The half-life of Pu-238 is 87.7 years, therefore between 1997 and the end of the Equinox mission in 2010 (13 years) one would simplistically expect the power output to fall by a factor of $\exp(-13/87.7)=0.86$ i.e. from 880W at launch to 758W. However, the thermoelectric converters would also degrade over time, in part owing to thermally activated migration

of the semiconductor material. In flight, the degradation has been modest and power output was expected to remain above 600W until the end of the mission in 2017.

An RTG is a unique and very special item. Whilst the Pu-238 fuel is not fissile (meaning that it cannot make an atomic bomb), it is radioactive and poisonous, and is classified as Special Nuclear Material by the US Department of Energy. The Pu-238 used on Cassini (about 82% Pu-238, with small amounts of the other isotopes) was manufactured by irradiating neptunium-237 with neutrons in a reactor at the DoE Savannah River site in Georgia, and was encapsulated in components made by the Oak Ridge National Laboratory in Tennessee. The encapsulation process had to be performed in a glovebox environment at Technical Area 55 of the Los Alamos National Laboratory in New Mexico. The encapsulated fuel was then integrated with the generator itself (supplied by Lockheed Martin in Valley Forge, Pennsylvania) at the DoE Mound Laboratory in Ohio, prior to being shipped to Florida.

The National Nuclear Security Administration requires that Special Nuclear Material be conveyed by the Office of Secure Transportation. This uses unmarked trucks protected by heavily armed guards (typically former members of the Special Forces) who are authorised to use deadly force to protect their hazardous and exceptionally expensive cargo.

Launch of RTGs requires a specific authorisation from the White House, and following a detailed review of safety analyses and the Cassini Environmental Impact Statement, approval was granted finally on 3 October 1997.

The GPHS-RTG (General Purpose Heat Source) built on the successful silicon-germanium alloy unicouple heritage of the Multi-Hundred Watt Radioisotope Thermoelectric Generators (MHW-RTGs) was used on NASA's Voyager missions and the USAF LES-8 and LES-9 communications satellites in the Lincoln Experimental Satellite series.

The GPHS-RTG was 1.14m long and 0.422m in diameter. The average mass of the three RTGs made for Cassini was 56.42kg (slightly heavier than those for Galileo, due to a change in the pressure relief device and

RIGHT Internal structure of the GPHS-RTG. The centre is the GPHS blocks. Heat flows radially out through the array of unicouples to the external structure. There the heat is removed (on the ground) by water-glycol coolant circulated through tubes, or (in space) by radiator fins. An inert atmosphere is held inside the unit on the ground, and is evacuated into space automatically on launch by a pressure relief device. *(NASA/DoE)*

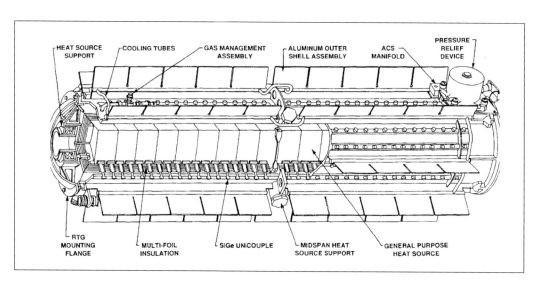

RIGHT 'No user-serviceable parts inside.' The GPHS block securely holds the fuel clads, with each iridium-coated fuel pellet releasing 62.5W of thermal power. The blocks were designed to tolerate a launch explosion and the heat of hypersonic entry from space. *(NASA/DoE)*

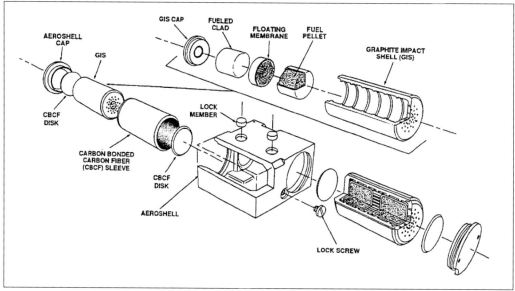

RIGHT The silicon-germanium unicouple includes a stack of semiconductor plates which convert a portion of the heat flow from the hot shoe at the bottom to the shunt at the top into an electrical current. The materials must tolerate operation for decades at very high temperatures. *(NASA/DoE)*

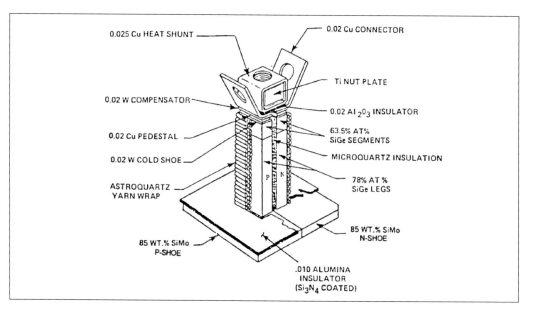

adapter plate). The structure was machined from a forging of 2219 aluminium alloy, and was coated in a specially developed high-emissivity silicone paint called PD-224.

The GPHS module was a graphite block that was designed to survive a launch explosion, or the more demanding instance of atmospheric entry from space. It contained the four fuel pellets (also known as 'clads'), each of which was a welded capsule made from iridium, which was resistant to corrosion even above 2,000°C. If any man-made artefact can be considered to be all but indestructible, then this is it.

The GPHS assembly consisted of 18 GPHS modules, each containing four fuelled clads which, in total, with 'fresh' fuel, yielded about 245W of heat per module for a total of 4,410W. The converter contained 572 silicon-germanium alloy thermoelectric elements ('unicouples') which converted this thermal power into at least 285W of electricity at the beginning of the mission.

The unicouples were held under a protective argon atmosphere on the ground – this gas helping to keep the unicouples cool and inhibit sublimation. But the gas would reduce efficiency in space, so the PRD (a spring-loaded bellows which pushed a lance through a sealing diaphragm as the ambient pressure fell off) allowed it to vent during launch.

The high-temperature operation (the unicouple roots were at ~1,000°C; the case temperature was about 250°C in space) and the radiation meant very carefully selected materials were needed. Molybdenum foil and Astroquartz fibre were used as thermal insulation and the unicouples were coated in silicon nitride to retard sublimation. Zirconia and alumina were used as electrical insulators and some parts were made of tungsten or Inconel (a nickel-chromium alloy). The materials and current loops in the RTG, coupled with the sensitive demands of Cassini's magnetometer, meant that the magnetic field produced by each RTG had to be measured and reduced by compensating magnets.

In addition to the RTGs that powered Cassini, there were 117 Radioisotope Heater Units (RHU). These were pellets of clad fuel in a cylindrical carbon aeroshell (like the GPHS) for safety, that each provided about 1W of heat, and could usefully maintain

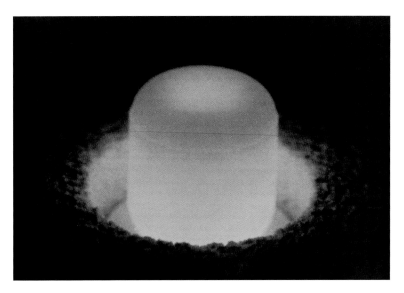

ABOVE A pellet of plutonium dioxide fuel emits an orange glow from the heat of natural radioactive decay. Normally in air the pellet is not actually red-hot: this picture was taken by insulating the pellet under a graphite blanket for few minutes to heat up, removing the blanket, and taking the picture before it could cool down. *(DoE)*

BELOW An environmental health specialist checks the radiation level near a fuelled Cassini RTG at the launch site in Florida. No protective clothing was needed. The RTG was sitting on a special handling pallet which included the lower part of the shipping cask in which it was sealed under an inert gas for transportation. Two small silvery flexible tubes are visible connecting the RTG to the cask base – these carry water/glycol coolant to help keep temperatures low for most of the time on the ground. *(NASA/KSC)*

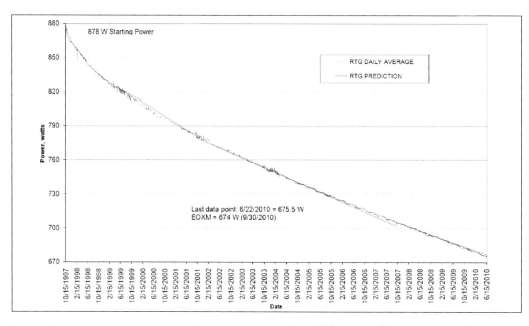

benign temperatures on systems which would otherwise require continuous electrical heating. This ability to provide continuous heat was particularly important on Huygens (which had 35 RHUs) for the lengthy coast after release, when it would be essentially unpowered, 10 AU from the Sun.

Yet even with three RTGs (probably the useful limit of RTG power; for more than 1kW in deep space it may be more practicable to use fission power) there was simply not enough power to operate everything simultaneously, although for the descoped Cassini without scan platforms the different pointing demands of the various instruments introduced incompatibilities anyway. So one mode might operate the RADAR and INMS instruments along with the plasma instruments, and use the thrusters to point the spacecraft. In this case the optical instruments would be placed in sleep mode and the RWAs would be powered down. Another mode might operate the optical remote sensing (ORS) instruments with the RWAs for precision pointing but with the thrusters and RADAR off.

As the Cassini mission was extended beyond the end of the prime mission in 2008, and the RTG power had slowly declined, some margins were clawed back in order to 'keep the lights on', but towards the end of the final mission extension some restrictions had to be imposed to exclude simultaneous operation of certain instruments.

The 30V output of the thermoelectric system fed a balanced power bus, with the RTG casings isolated from the rest of the spacecraft chassis. In principle this balanced bus should have been ±15V relative to the chassis, but changing short circuit conditions caused gradual and sometimes stepwise changes in the voltage.

The RTGs played a part in a deep space mystery known as the so-called Pioneer Anomaly. Essentially, the tracking data for deep space probes (in particular Pioneer 10) indicated a slight deviation from the trajectory expected from classical gravity. The implications of a deviation from the inverse square law would be profound. However, careful analysis indicated in 2012 that the recoil from thermal photons radiated away from the warm surfaces of the RTGs explained most or all of this discrepancy.

Message from Earth

A creative project to have Cassini and Huygens carry a 'message from humanity' in a similar manner to the famous 'Voyager disk' had intended to install a small diamond disk from DeBeers, 2.8cm in diameter and 1mm thick.

The diamond was to be microscopically etched using an oxygen plasma to write (in 64 shades of grey) a message which included an astronomically accurate portrayal of the solar system and Cassini's trajectory through it, a specially arranged picture of a family of Earthlings on a beach on Hawaii, one of whom was holding the disk for scale, and renderings of Earth from which readers of the disk in the far future might be able to deduce (from the progress of continental drift) when it had been sent.

Sadly, concerns about commercial sponsorship of the project, and squabbles over who was to take credit, led to its abandonment. The diamonds never flew.

Instead, a DVD was burned with 616,403 signatures. As a means of communicating with extraterrestrials (or, for that matter, humanity in the far future) a DVD is a rather imperfect device because not only might its thin metallisation corrode away, but the compressed digital format of its microscopically-written data will be inscrutably challenging to interpret, even if it can be physically read. However, perhaps that isn't so important. The DVD project (and a separate CD-ROM on Huygens) gave a way for ordinary people – taxpayers after all, who made Cassini possible – to feel connected to the mission. Indeed, a number of later missions have undertaken similar exercises to engage the public.

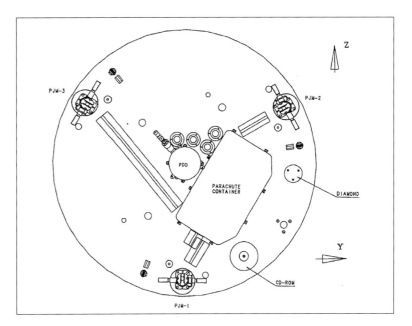

ABOVE The diamond disk never flew, but engineering drawings (this view is the Huygens probe top platform) showed where it would have been accommodated; also labelled are various parachute system components and a CD-ROM of messages from the public. *(ESA/Aerospatiale)*

BELOW LEFT Installation of the DVD (DVDs became available only in 1995) carrying over half a million signatures in digital form onto Cassini. The white CIRS cooler is visible at the top, and the black thermal cover for an RTG is near the bottom. *(NASA/KSC)*

BELOW The last items to be designed for a spacecraft are the balance weights that will set the geometry of the spin axis and moments of inertia after all the other masses, like instruments, have been delivered. In an informal gesture not uncommon in the space world, the Huygens spin balance weights were engraved with the names of some key project personnel. *(P. Couzin)*

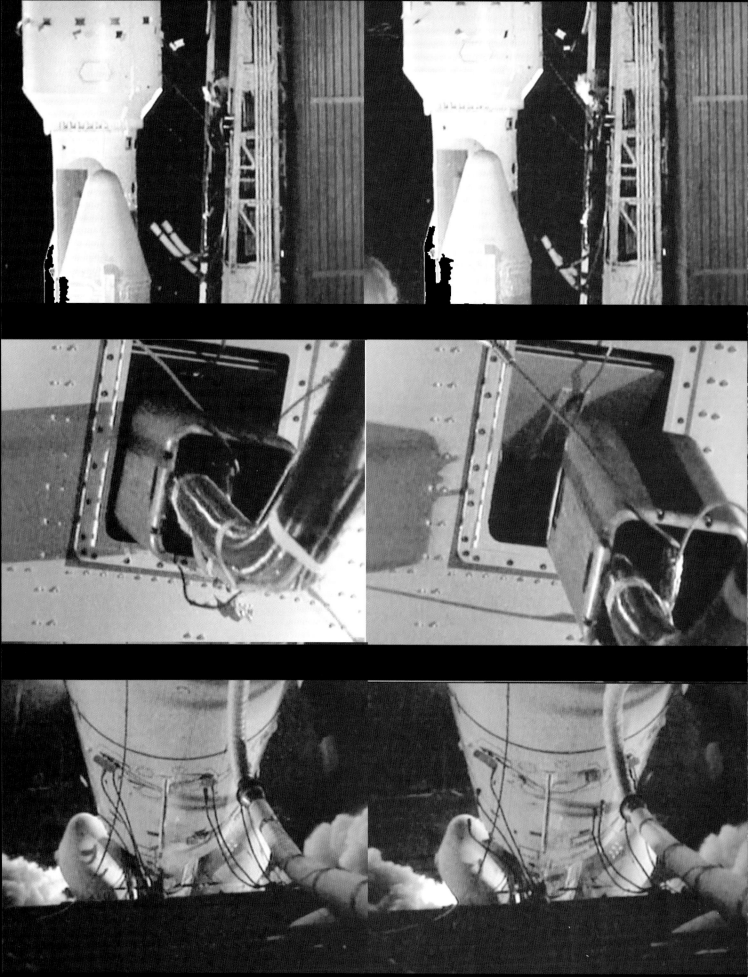

Chapter Five

Launch and cruise

—●—

After a thunderous night-time ascent into space, Cassini's epic voyage of exploration around Saturn would be preceded by a seven-year solar system tour that was epic in its own right. This long trip would give time to design the detailed trajectory at Saturn, plan the myriad observations, and shake down problems with systems and instruments.

OPPOSITE The moment of lift-off is a delicately choreographed series of events, with dozens of umbilical cables and hoses separating and covers being released or swinging into position. In this case these events were monitored by Air Force cameras on the launch pad. *(Stephen Slater/USAF)*

Launch at last!

The hiatus caused by the Huygens cooling muddle pushed the launch back by a week, to 13 October 1997. Hundreds of scientists and engineers, in many cases with their families, gathered in Cocoa Beach, Florida for the launch, marking for many the culmination of over five years of effort in design, construction, assembly and test. The event would, it was hoped, be a cathartic spectacle – the climactic reward for years of lost weekends and evenings. For many engineers, this would mark the end of their involvement. For some of the scientists, especially on the probe, it would be just the beginning of a long wait. For everyone, it was a transition.

It was also a worry. Rocket reliability had improved over the space age, but it was a simple fact that no launcher by the end of the 20th century did much better than 98% or so. The odds were that if anything were to go badly wrong with Cassini, it would happen at launch.

The Titan IV had a reasonably good record, with four successful launches per year of military satellites between 1994–96, although memories were still fresh of a spectacular 1993 explosion. Cassini was the only non-military payload assigned to this rocket. Furthermore, this was to be only the second launch of the Titan IVB variant, using the new SRMU solid rocket motors. The first had successfully launched a military early warning satellite only in February 1997.

Another factor added tension to the proceedings. Legal challenges had been initiated to prevent Cassini's launch, as part of some bizarre crusade against nuclear power in space. There were genuine concerns of course, but in part perhaps "Halt nuclearization [sic] of space: Stop Cassini" just made a good cause for those who needed something to protest about in the decadent dot-com world, between the end of the Cold War and before the tragic events of 9/11. The White House granted the nuclear launch safety approval on 3 October. An attempted injunction in a Hawaii District Court to stop the launch was dismissed. A group of 27 activists who crossed the fence at Cape Canaveral Air Force Station were arrested and charged with trespassing on a government facility.

In the foetid early morning darkness of 13 October, hundreds of scientists and engineers disembarked the buses delivering them to the Carrs Park viewing site, a few miles from the launch pad known as SLC-40. (The only other interplanetary launch to depart from this pad was the ill-fated Mars Observer, which was on a Titan III in 1992.) They nervously found places on bleachers and settled down to watch. The 140min launch window opened at 04:55hrs.

But the launch was scrubbed due to excessive winds at altitude that could stress the rocket during ascent. Everyone trudged back to their hotels to get some sleep before science planning meetings later in the day. The launch was rescheduled for two days later. The window remained 140min but would advance by about 6min/day.

This time, the countdown proceeded without hiccup and at 04:43hrs (Eastern time) on 15 October 1997 the ignition command was sent. Observers (including the author) saw flames erupt silently from the base of the rocket – an initially unsettling spectacle. But a second or so later, a rumbling roar could be felt and heard, transmitted through the water between the pad and the bleachers. The rocket began to climb, on a searing pillar of flame. It was already a rocket's length above the pad before the direct sound reached us, a wall of deep crackling bass, barely challenged by the cheers of the crowd.

Although the sky overall was clear, as launch safety rules demanded, there was a single low cloud that had drifted almost ominously, over the pad. Some feared that it might force a launch scrub, but it was a small cloud with no precipitation or lightning, and hence was not of concern. But as the Titan climbed through it, the cloud lit up like a Chinese lantern, illuminated from within by the brilliant exhaust from the two SRMUs. It seemed like a good omen.

The three-segment SRMUs were 3.2m (10.5ft) in diameter and 34.2m (112.4ft) in length. Each motor contained 312,458kg (688,853lb) of aluminium/ammonium perchlorate/polybutadiene solid propellant and provided a maximum thrust of 7.56MN (1.7 million lb) at sea level. Flight control was achieved by directing the thrust through a gimballed nozzle that was controlled by hydraulic actuators.

ABOVE A launch pad camera view of the bottom of the ascending Titan. The nozzles of the two liquid fuel engines of the core stage are lit by the dazzling luminance of the solid motor exhaust. The core engines are not ignited until 132 seconds after launch, just before solid burnout. *(Stephen Slater/USAF)*

ABOVE The Titan ascends on the brilliant exhaust from the two SRMUs. Note the US Air Force logo and how much wider the payload fairing is than the core stage. The tower at right is one of four that suspends steel cables around and above the pad to attract lightning away from the launcher itself. *(NASA)*

LEFT Views of Cassini's Titan IVB in ascent. Top: the vehicle climbs towards a cloud over the pad. Middle: following an initially vertical ascent, the vehicle pitches over in order to accelerate into orbit. The bottom of the fairing is faintly visible, illuminated by the glow of the exhaust from the solid boosters. Bottom: the two main engines of the core vehicle are seen as a glowing spot in the centre. After the solid motors separate and tumble, some fiery exhaust still spews from their nozzles. The faint bright spot near the top of each booster is the glow of their separation motors. *(Stephen Slater/USAF)*

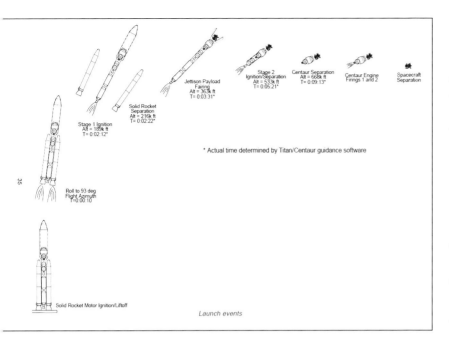

Stage 1 Ignition
Alt = 189k ft
T = 0:02:12*

Solid Rocket
Separation
Alt = 216k ft
T = 0:02:22*

Jettison Payload
Fairing
Alt = 363k ft
T = 0:03:31*

Stage 2
Ignition/Separation
Alt = 533k ft
T = 0:05:21*

Centaur Separation
Alt = 668k ft
T = 0:09:13*

Centaur Engine
Firings 1 and 2

Spacecraft
Separation

Roll to 93 deg
Flight Azimuth
T = 0:00:10

* Actual time determined by Titan/Centaur guidance software

35

Solid Rocket Motor Ignition/Liftoff

Launch events

**ABOVE The Titan
IVB-Centaur launch
sequence.** *(NASA)*

As the rocket continued to ascend, it began to arc out over the Atlantic Ocean. After 2min 11sec, the first stage liquid engines were ignited, virtually invisible against the glare from the SRMUs (the metal powder in the solid propellants gives them their dazzling luminosity, and the resulting metal oxide exhaust leaves a persistent thick smoke trail). The solids burned out a few seconds later at an altitude of 68km. A set of six staging rockets on each SRMU ensured positive separation from the core vehicle, which continued to thrust. The spent motors tumbled away to splash into the ocean.

Three and a half minutes after launch, the

aluminium payload fairing, 5.9m (16.7ft) in diameter and 20m (66ft) long, was jettisoned at 120km altitude where the air was sufficiently thin that aerodynamic stresses and heating were no longer of concern and the fairing had become just dead weight. After another 2min, the core stage was depleted and the second stage separated and ignited at an altitude of 168km, to insert the Centaur stage and its Cassini-Huygens payload into orbit 3.5min later. Separation was achieved by firing four retrorockets and igniting a SuperZip pyrotechnic charge that severed the linkage between the Titan and the Centaur.

The Centaur stage, whose development was managed by NASA's Lewis Research Center and was made by Lockheed Martin Astronautics, was 8.9m (29.45ft) long and 4m (14ft) wide. It provided a thrust of 147,000N (33,000lb) using two RL-10 engines that burned cryogenic liquid hydrogen and liquid oxygen. These ultra-cold volatile propellants delivered the highest performance, which is most strongly leveraged on the upper stages of a rocket.

At Titan-Centaur separation, flight control transferred to the Centaur, which started a pre-burn sequence, then ignited for the first of two burns. The first burn lasted 131sec. The Centaur-spacecraft combination coasted in parking orbit until the trajectory lined up with the direction for interplanetary injection, about 15min later. The parking orbit had a perigee (closest point to Earth) of 170km and an

TITAN IVB PROPULSION

Stage	0	1	2	3
# Engines	2 USRM	2 LR-87	1 LR-91	2 RL-10
Thrust (kN)	15120	2440	467	147
Specific Impulse (s)	286	302	316	444
Propellant*	Solid	N_2O_4/A-50	N_2O_4/A-50	LO_2/LH_2
Burn Time (s)	140	164	223	625

The Hercules Upgraded Solid Rocket Motor (USRM – also referred to as SRMU) burned a solid propellant composed of ammonium perchlorate (NH_4ClO_4) and aluminium powder, bound with a synthetic rubber called hydroxyl terminated polybutadiene (HTPB). The liquid propellant first and second stages of the Titan IV used the hypergolic (ignite on contact) propellants dinitrogen tetroxide (N_2O_4) and Aerozine-50, the latter being an equal mix of hydrazine (N_2H_4) and unsymmetrical dimethyl hydrazine ($H_2N_2(CH_3)_2$, abbreviated as UDMH). The second stage produced a slightly greater specific impulse (thrust per unit propellant mass flow) because by operating at higher altitude it was able to employ an engine nozzle that flared out more for improved efficiency

apogee of 445km, and was designed to provide an orbital lifetime of about 20 days just in case the Centaur failed to restart its main engine successfully and time was needed to overcome the problem.

Happily, the Centaur reoriented itself using hydrazine thrusters as planned and reignited its main engines for the 8min burn required to accelerate out of Earth orbit and into solar orbit. It receded from Earth with a launch energy (C3) of 17km^2/s^2 corresponding to a departure velocity of about 4km/s.

The Centaur adopted an attitude in which the HGA on the far end of Cassini was aimed at the Sun, then commanded Cassini to fire the separation pyrotechnics. In less than 6.5min from the Centaur completing its second burn, Cassini was now flying freely at the start of what would become two decades in space.

About 20min later, after an interval calculated to allow the Cassini spacecraft and the Centaur to drift apart to avoid impingement of exhaust from the Centaur's thrusters onto Cassini, the spent stage executed a final manoeuvre in order to avoid impact with Cassini or Venus, while controllers using the 34m DSN dish in Canberra, Australia established contact with the Sun-pointed Cassini through a Low-Gain Antenna.

After initial systems checks, controllers executed 25 days after launch the first Trajectory Correction Manoeuvre (TCM-1), the first of 21 burns planned on the way to Saturn. This 35sec burn of Cassini's main engine nudged the spacecraft by 2.7m/s, in part to correct for the fact that launch occurred right at the beginning of the window on 15 October, rather than 40min later.

Instrument checkouts soon began. Protective covers on the optical instruments were released. The three antennas for the RPWS instrument were deployed. Software updates and equipment configurations were made. By the end of the year, Cassini was more or less fully checked out.

Ironically, the delay due to the Huygens cooling incident may have in some ways been a blessing. By forcing the launch more towards the middle of the launch window, instead of the very beginning, Cassini's trajectory was closer to the optimum, such that some on-board propellant reserves were not needed. This

margin would prove essential in permitting a later redesign, and subsequent extension, of the Cassini tour at Saturn.

Paradoxically, Cassini-Huygens did not immediately head for the outer solar system but went inward toward Venus, heading for the first of several planetary flybys that would augment its orbital energy to enable it to climb out of the Sun's gravitational potential well to reach Saturn. These flybys offered the Cassini program opportunities to develop coordinated observation sequences to be undertaken during a planetary encounter.

At the time of launch, the plan was for Cassini to undertake very little activity prior to its approach to Saturn. Consequently, science observations were minimal during the first encounter with Venus on 26 April 1998, just six months into the mission. The Radio and Plasma Wave Science instrument (RPWS) 'listened' for, but found no evidence of radio emissions from possible lightning. These early sequence designs had to be done by hand or with rudimentary software tools, as the development of software for automation in spacecraft control and planning which would be essential for undertaking the mission at Saturn had been deferred until after launch.

While in the inner solar system, the 4m dish antenna that was aligned with the spacecraft's –Z axis was generally pointed toward the Sun in order to shade the rest of the spacecraft. Although there was some freedom to roll about this axis, typically the Huygens probe (along the –X axis) was kept in the ecliptic and in the

BELOW A display at NASA's Glenn Research Center of the Centaur G-Prime stage which was originally to have boosted Cassini after its deployment in parking orbit by a Space Shuttle. The spherical helium pressurant tanks and the nozzles of the two RL-10 engines are visible to the right. *(Author)*

direction of the spacecraft velocity to shield against micrometeoroids. The orientation of the two secondary antennas with respect to Earth was also important, and all attitudes had to be examined closely for thermal effects, especially with regard to the sensitive radiators for the Visual and Infrared Mapping Spectrometer (VIMS) and Composite Infrared Spectrometer (CIRS). It was also necessary to prevent light from scattering off parts of the structure onto the two Stellar Reference Units (SRU) that provided data for attitude control. These factors restricted pointing for science observations.

The second flyby of Venus occurred at an altitude of 598km at 20:30:07 UTC on 24 June 1999, with the spacecraft making its approach from the dusk-side of the planet. This provided a unique opportunity for the fields and particles instruments to study how the plasma of the solar wind interacted with Venus, a planet that possessed no intrinsic magnetic field. Cassini had a much more capable fields and particles payload than previous spacecraft to encounter the planet.

Since the flyby was at a low altitude, the spacecraft could directly investigate Venus's ionosphere. The CAPS, MAG, MIMI and RPWS instruments were all powered on several hours before closest approach. RPWS measured plasma waves and repeated the lightning search, while the MIMI-CHEMS sensor made measurements in the ionosheath and stagnation region at Venus, detecting a number of energetic species, possibly including O+ and C+ ions 'picked up' from the atmosphere. MIMI was able to detect energetic neutral atoms escaping from the atmosphere with its Ion and Neutral Camera (INCA).

The magnetometer remained stowed in its canister because Cassini was still too close to the Sun to expose the boom material to solar heating. However, the magnetometer sensors themselves could still be operated (albeit with a stronger background from the spacecraft) and they detected the crossings of the bow shock.

Several hours prior to closest approach, Cassini executed a 34° roll about the Z axis in order to point the ORS instruments at the planet as it swept across the night-side, and VIMS reported the first detection of thermal emission from the Venusian surface at 0.85µm and 0.90µm.

Four trajectory correction manoeuvres were executed during the 54 days between the Venus-2 and Earth flybys, with the final correction being made seven days before reaching Earth. As a safety measure (given the presence of RTGs), the Earth-avoidance strategy had Cassini initially aimed far away from Earth so that a collision was impossible in the event that control was lost, but the miss-distance was gradually reduced as the trajectory measurements improved, to achieve the correct distance to put it on the proper trajectory to Jupiter.

In preparation for making observations during the Earth encounter, the VIMS infrared-radiator and optics covers were released two days prior to closest approach. Then, on 16 August, the 11m magnetometer boom was unlatched. It was essential to deploy the magnetometers into their operational configuration during this flyby in order to obtain calibration data in a relatively well-known and strong magnetic field. Elastic energy in the boom's longerons drove their deployment, with a viscous rate limiter preventing the boom from springing out too rapidly. Cassini's rapid dash through Earth's magnetosphere provided a detailed snapshot for comparison with various Earth-orbiting satellites.

At 21:28 UTC on the day before the closest approach, the spacecraft was rolled 104.3° about the Z axis. This permitted the Moon to traverse the fields of view of the ORS instruments. For the Venus flybys the deadband for pointing had been a sloppy 20mrad, but this was improved to 2mrad for the Moon, which was observed at quarter phase by ISS, UVIS and VIMS for about 29min. The VIMS data indicated that the lunar regolith contained water. This was a surprising result because the detection of water in the Apollo samples had been assumed to be terrestrial contamination. This detection was later confirmed by other spacecraft.

The closest approach to Earth occurred at 03:28 UT on 18 August 1999 at an altitude of 1,163km, while passing over the southern Pacific Ocean. For several minutes the geometry was such that, with Cassini still using its HGA as a Sunshield, the Earth drifted into the field of view of the antenna. This permitted the RADAR to transmit and receive pulses along a track that began over the Pacific and extended across South America, marking the first time that the radar could be tested end-to-end. Such a test hadn't been possible at Venus because its dense atmosphere was too thick for the instrument to get an echo from the surface.

A post-Earth trajectory change manoeuvre 13 days after the flyby set Cassini on its proper course to pass Jupiter on 30 December 2000.

On 23 January 2000, Cassini briefly observed asteroid 2685 Masursky. Although the distance of some 1.6 million km meant images showed no detail, they indicated the body to have a diameter of 15–20km.

Jupiter flyby

Unlike the Voyagers, which ventured deep into the Jovian system, Cassini's relatively distant flyby of the giant planet provided only a modest gravitational nudge. In a Jovicentric view, the Cassini trajectory was almost a straight line. However, the encounter contributed to Cassini's heliocentric energy and enabled it to reach Saturn.

The Jupiter encounter gave scientists an important opportunity to shake down their instruments and test the software which would be used to plan observational programs at Saturn.

Cassini's observations of the Jovian system spanned a period of six months, beginning on 1 October 2000. It flew by the planet on 30 December at a range of 137 Jovian radii (R_J), which was 9,794,130km above the cloudtops.

Scientifically, even though the flyby was distant (precluding any ground-breaking remote sensing of the Jovian moons, for example), the long period of observation provided an opportunity to help to recover some of the science goals observing Jupiter's atmospheric dynamics that had been denied to the Galileo mission when its flexible High-Gain Antenna had failed to deploy properly, curtailing the rate at which it could transmit data to Earth by a factor of many thousands. In particular, data-intensive cloud-tracking movies were now feasible.

Moreover, Galileo was still operating, and the opportunity to study the dynamic magnetosphere at two locations simultaneously would improve understanding of the interaction of the planet's magnetic field, the solar wind, and the trapped plasma in Jupiter's powerful and deadly radiation belts. The observations by

the two spacecraft in-situ were coordinated with ground-based telescopes, including the Very Large Array (VLA) radio facility in New Mexico, and with both the Hubble Space Telescope and the Chandra X-ray Observatory in Earth orbit.

The planned collaborative Cassini-Galileo observations spanned 100 days, beginning on 26 October. Galileo collected data continuously as it travelled from the solar wind through the bow shock and the magnetopause into the middle and inner magnetosphere, and then back through these regions into the solar wind again downstream. Meanwhile, RPWS monitored Jovian radio emissions, and plasma and radio-wave phenomena associated with the bow shock.

One of the more remarkable observations of the magnetosphere during this period was by Cassini's radar instrument. Although much too far away to perform active radar measurements, its dish receiver could be used as a passive radio telescope, measuring the synchrotron emission produced by electrons in Jupiter's strong magnetic field. Cassini's measurements

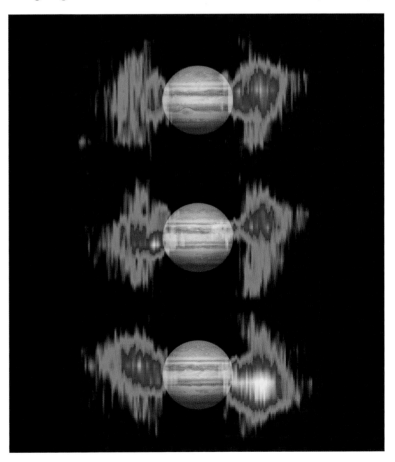

BELOW Jupiter's radiation belts, revealed through synchrotron emissions mapped out by Cassini's radar receiver operating as a microwave radiometer. The magnetic field is inclined to the equator, making the radiation belts appear to 'wobble'. *(NASA/JPL)*

at a wavelength of 2cm were coordinated with those obtained by the VLA at 20cm and 90cm. The spacecraft was slewed around in a raster pattern twice – rotating the spacecraft about the Z axis between each $8R_J$-wide image to change the polarisation of the measurement. Combining this polarimetric 2cm image (with a resolution of ~$0.25R_J$) with ground-based data provided the best estimate to date of the energy spectrum (up to 50MeV) and density of the electrons circulating in the radiation belts. Observations of synchrotron emission provide the only 'remote probe' of high-energy electrons in Jupiter's inner radiation belts.

Inbound to Jupiter, Cassini monitored the solar wind while Galileo measured magnetospheric properties and the Hubble Space Telescope imaged the aurora. Outbound, Cassini monitored the night-side aurora while Galileo monitored the solar wind and Hubble observed the day-side aurora, showing the influence of solar wind on the aurora.

In fact, fluctuations in solar wind pressure and the rotation of the Jovian magnetic field caused the bow shock and the magnetopause to pulsate while Cassini was inbound. Indeed, no fewer than 44 bow shock crossings and six crossings of the magnetopause were evident in data from the magnetometer and CAPS.

The sensitivity and mass resolution of CAPS permitted the first detailed compositional analysis of the thermal plasma in the outer regions of Jupiter's magnetosphere – identifying ion species associated with oxygen, sulphur, potassium and related molecules. Meanwhile, higher-energy particles were measured by MIMI.

During the period from 90 days to 20 days prior to the encounter, the ORS instruments observed all odd rotations of Jupiter, plus one even rotation every 120hr. CIRS mapped the planet's thermal structure and the aerosol loading of the stratosphere, and searched for new spectral features. ISS and VIMS observed every 60° of longitude in order to map the life cycles of storms and jets and measure wind speeds. This planned movie sequence was interrupted when the attitude control system aborted science operations in response to an out-of-limit torque indication from one of the reaction wheels.

Although Cassini's spatial resolution couldn't compete with that of Galileo, its spectral range was rather better and Cassini had the opportunity to observe the four large moons and the thin and faint Jovian rings at very low phase angles (i.e. with the Sun almost directly behind Cassini), allowing measurement of the 'opposition effect' that informs the size and texture of the ring particles and the satellite regoliths. (As an example, our own Moon is 20 times brighter when it is full than it is when at half phase. A factor of two results from the increase in illuminated portion of the disk, but 10 times is attributable to the decrease in mutual shadowing of lunar regolith particles.) Cassini also observed one of Jupiter's smaller (~160km), distant satellites, Himalia, measuring how its brightness changed with rotation to deduce its shape, and measuring its reflectance spectrum with ISS and VIMS respectively to characterise its surface material.

ISS observed the volcanic moon Io while that was passing through Jupiter's shadow in order to examine the morphology and temporal variability of eruptive plumes, atmospheric airglow, and surface hot spots. It also studied changes in Io's optical emission due to changing interactions between the magnetosphere and the satellite's thin exosphere. UVIS observed changes in the torus of ionised gas which occupies Io's orbit.

BELOW Although the Jupiter flyby was relatively distant, Cassini's camera acquired some astonishing images, such as this one of Io viewed close to Jupiter's terminator. *(NASA/JPL/SSI)*

During the years of cruising through interplanetary space, Cassini's Dust Analyser measured the dust population, bringing the new capability of measuring the particles' composition. It followed up on the discovery by the Ulysses mission in 1992 that narrow dust streams emanated from Jupiter – which the Galileo mission later traced to Io. These fast but tiny particles are charged by Jupiter's magnetospheric plasma, allowing the fast-whirling magnetic field of the planet to fling them outwards. Encountering these streams around 29 December, CDA revealed that the particles were blasted out at 300–400km/s, which was much faster than had been expected.

Cassini continued to observe beyond Jupiter, with its outbound movie sequence showing a thin bright crescent. It observed every odd rotation of Jupiter, plus one even rotation every 120hr, from +20 to +90 days. Observations of lightning on the night-time hemisphere showed the flashes were associated with storms in the cyclonic belts of the planet's circulation.

Although Cassini's ISS performed well at Jupiter, planners got a scare in 2001 when the images became hazy, perhaps as a result of thruster exhaust or some other contamination being deposited on the optics or detector. Fortunately, the issue was resolved by cycles of careful heating over a period of several months to 'burn off' the contamination.

In December 2001, while travelling through the quiet emptiness between Jupiter and Saturn, Cassini's radio link was used to search for astrophysical gravitational waves. These ripples in space-time were predicted as a consequence of general relativity, and if large enough would be detectable as a transient distance change that would be measureable by monitoring the phase of Cassini's radio signal as it was received by the DSN antennas.

Although no gravitational waves were detected, the spacecraft's combination of two-

ABOVE Cassini ISS imaged stars for calibration purposes. On the left is HD339457, taken on 25 May 2001 before any problems were noted. The second image was taken of Maia in the Pleiades on 30 May and has a blurry ring due to some kind of contamination. Two week-long heat treatments (at –7°C and +4°C) improved the third and fourth images (of Spica) somewhat. Further treatments at 4°C lasting 4 months finally restored Spica to a satisfactory sharpness in July 2002 (right). *(NASA/JPL)*

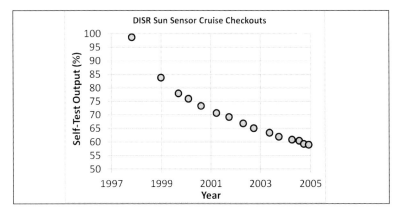

way X-band and Ka-band tracking was the most sensitive search for this phenomenon up to that time. Unfortunately the two-way Ka-band system failed before further experiments could be performed to test general relativity. (The first confirmed detection of gravitational waves was made in 2015 by ground-based instruments, 99 years after their prediction.)

Although originally justified to budget planners as an operations rehearsal (in which respect it succeeded and was important, e.g. in shaking down the reaction wheel behaviour as a critical issue) the Jupiter flyby was a significant scientific bonanza.

Probe checkouts continued at six-monthly intervals to monitor any changes in sensitivity of the instruments (wherever possible, the instruments included some sort of self-test stimulus). Generally minor drifts were observed in zero-offsets or readings (e.g. some channels of DISR lost 1% or so of signal, probably due to yellowing of the fibre optic bundle), but there was a drop of several tens of per cent in the self-test

ABOVE During the seven years that Cassini spent in space the self-test signal level of the DISR Sun sensor of the Huygens probe dropped by 58% over the time of the 16 in-flight checkouts. It is not known whether this was due to a decline in the output of the stimulus light-emitting diode or to a drop in the sensitivity of the photodiode detector, or indeed both, probably as a result of displacement damage from neutrons emitted by the RHUs and RTGs. *(Author)*

signal of the DISR Sun sensor: this sort of change underscored the value of health monitoring on long cruises to the outer solar system.

The seven-year cruise to Saturn posed challenges not just to the hardware in space but also to the integrity of the mission's ground systems, and the retention of knowledge by personnel. As there was no way to pay the development teams for years of quiet cruise, many skilled staff moved on to other projects or left the field entirely. And even the most impressive observation-design software was worthless if an upgrade to the computer operating system made obsolete the graphics card upon which it relied to operate. Careful software maintenance, detailed documentation, and creative staff management are necessary to preserve expertise and capability on outer solar system missions.

Propulsion system regulator trouble

Cassini had the most elaborate propulsion system yet flown on a deep space mission, in part to perform the formidable velocity changes that would be needed to reach Saturn and then to enter orbit around the planet, but also because the mission would last so long that reliability had to be ensured with redundant systems. Mission controllers had to call upon this resilience rather earlier than they might have hoped.

The main engine (or rather, whichever of the two redundant engines was chosen for a given burn) used hypergolic monomethyl hydrazine and dinitrogen tetroxide. The tank pressure had to be high enough to force the propellant cleanly into the engine without 'chugging', but obviously the need to make the large tanks lightweight meant their maximum pressure was limited. To sustain the pressure at about 225psi as the propellant was used, a regulator was to adjust the flow of helium (stored in its own tank at an initial pressure of 954 psi) to pressurise the propellant tanks.

The regulator for Cassini used a valve having a hard seat, because at the time the system was designed the soft Teflon seat used on Galileo had been feared to be susceptible to seat extrusion by slow creep of the material.

However, a hard seat was susceptible to a higher leak rate if a particle of debris (e.g. from a pyrotechnic valve) were to get caught on the seat. Moreover, the filter that was sometimes used to protect the regulator against such particles was itself feared to be susceptible to oxidant vapour corrosion (a factor considered as a possible contributor to the loss of Mars Observer in 1993). Hence a backup regulator was included, with helium latch valves and a ladder of pyro-valves designed to isolate and bypass a failed component.

In preparation for TCM-1, Pyro Valve #1 was fired to open helium flow to the system, and latch valve LV-10 was commanded open to begin tank pressurisation via the regulator. However, once LV-10 was opened, it was observed that the pressure of the nearly full tanks continued to rise beyond the 245psi set point at an alarming rate. This indicated a leak through the regulator of some 1,700scc/min – a factor of 1,000 higher than the worst case expected from ground tests! A contingency command (thoughtfully already prepared just in case) was quickly sent from the ground to close LV-10 and staunch the leak.

It seems that firing the pyro valve may have released debris which prevented the valve seat from sealing. TCM-1 was conducted in 'blowdown' mode (letting the pressure fall without regulation as the fuel was used) without any further concerns. But when LV-10 was opened again to support the Deep Space Manoeuver, the regulator leak was found to have increased by a further factor of 6.6.

Fortunately LV-10 worked very well, with a much better-than-specified leak rate of its own. Opening it for brief periods allowed the tank pressures to be managed at a coarse level, but the pressure regulation and fault protection logic for Saturn Orbit Insertion had to be reprogrammed. However, the long cruise provided plenty of time for this recovery work, and the 'improvised' mode of operation has been problem-free.

In the event of LV-10 leaking (or failing open), Cassini's fault protection logic would kick in before the rising pressure could rupture the tanks and it would fire pyro valves to isolate the valve and regulator. At that point mission controllers would switch in the redundant

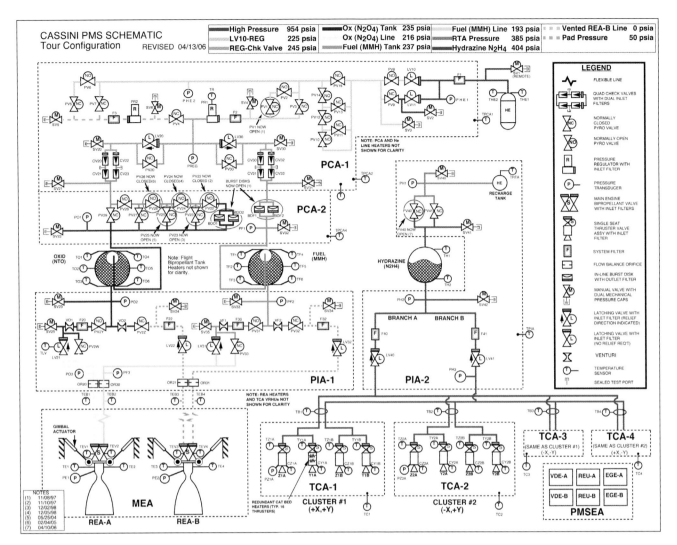

regulator. But this problem has never occurred, and by now there is enough empty ('ullage') space in the propellant tanks that the available helium is not enough to risk over-pressurisation in any case.

Designing the tour

A major task during the cruise to Saturn was to design Cassini's tour and plan the hundreds of thousands of observations, because the instruments would be useless without detailed schedules for their use.

For example, a camera can do useful science at Saturn, observing its cloud patterns from several million km away, but a magnetometer only measures the field at the spacecraft itself, and so can only shed light on Saturn's interior dynamo when it is within a couple of Saturn radii of the planet; i.e. about 120,000km. On the

other hand, a magnetometer functions all the time, whereas to track cloud patterns a camera needs to view the day-side of the planet. And of course, Saturn was not the only target; there were also Titan, the other satellites, the rings and the magnetosphere.

In practice, one doesn't design a mission like Cassini as a series of observing opportunities but as a path through space. Linking together a series of observations had to be done according to the laws of physics, and thus Cassini's path was fundamentally a series of elliptical (Keplerian) orbits around Saturn. Left to itself in such an orbit, the path would essentially repeat over and over – e.g. if it were an eight-day orbit that was inclined at 40° to the ring plane, Cassini would generally remain in an eight-day orbit at that inclination. But the observing opportunities for such a path would be quickly exhausted (or at least would be

ABOVE The forest of valves allowed Cassini's propulsion system to be reconfigured against failure. The prime regulator failed after launch and pressurisation was managed largely by latch valve LV-10. After the A-side thrusters degraded, in 2009 it was decided to switch the system to the B-side thrusters. *(NASA/JPL)*

limited in scope). Torquing the orbit around with rocket propulsion would be prohibitive in its fuel demands, so the trick was to employ the gravity of a moon to adjust the trajectory, much as the planets were used for gravity assists on the way to Saturn.

The Galileo spacecraft's tour of the Jovian system was able to exploit the four large moons, all of which were broadly similar in mass and orbited in the same plane, using their gravity to deflect its path around the planet in such a manner as to create a variety of observing opportunities. However, of the Saturnian satellites only Titan's gravity was strong enough to serve this purpose. On the other hand, Cassini's tour would not be confined to the equatorial (ring) plane, but would be three-dimensional.

The original 'proof-of-concept' tour for the joint mission study was designated 84-01 and it featured 34 Titan flybys over four years, beginning with the spacecraft's arrival at Saturn in February 2000. The tour included two Iapetus encounters and was to end in a highly inclined orbit that would enable Cassini to 'look down' on the ring system. Tour 84-02 tried to meet similar goals in only three years, while tour 85-01 (designed the following year and now anticipating arrival in 2002) explored whether it was possible for Titan's gravity to rotate the orbit 'petal' onto Saturn's day-side in order to

allow long imaging sequences. The bewildering range of possibilities, and the challenges of meeting the mutually conflicting science objectives, became readily apparent to the trajectory designers at JPL.

Another iteration (88-01) emerged for the Phase-A study. It featured 36 Titan flybys, as well as two of Iapetus and one each of Enceladus and Dione. Important considerations were also the inclination of the orbit (because a higher inclination would allow the spacecraft to look down on Saturn's poles, which are difficult to observe from Earth), and having the vehicle pass behind Saturn and Titan for solar and radio occultation observations.

During development through to launch, another tour (92-01) was the working plan. It was basically the 88-01 tour adjusted for the new launch date. These proof-of-concept tours, however, only scratched the surface of the colossal option space.

The range of possibilities can be thought of in the following manner. Apart from the very beginning of the tour (where the orbit derives directly from the arrival geometry and the insertion burn) and the very end (when the spacecraft must be disposed of safely), the orbit of Cassini must be (generally) resonant with the orbit of Titan. This ensures that the spacecraft will make periodic close encounters with Titan to use its gravity to hop from one resonant orbit to another; e.g. from a 1:2 resonance with the spacecraft in a 32-day orbit during which Titan goes around Saturn exactly twice, to a 3:4 resonance in which the vehicle makes three orbits in the 64 days it takes Titan to make four orbits. Depending on which side of Titan Cassini flies past, and how close the flyby, such encounters can increase the energy (and period) of the orbit or decrease it, thereby 'pumping' Cassini's orbit up or down. So far it has been assumed that the spacecraft travels in the plane of Titan's orbit and the closest point of approach occurs over the equatorial region. But if Cassini encounters Titan away from the equator, the encounter can change the inclination of the orbit too, 'cranking' it by up to 13°. By flying over Titan's northern hemisphere, the gravity will deflect the spacecraft's trajectory southwards, and vice versa. It follows then, that a given change to Cassini's trajectory also maps

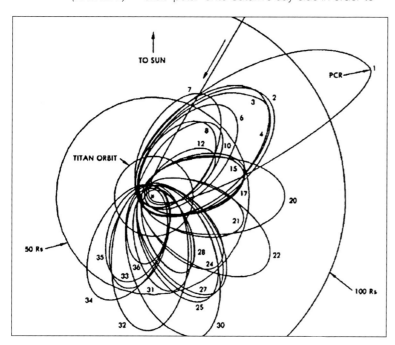

BELOW The early tour designs for Cassini (in this case, 88-01 developed for the Phase-A study) were essentially two-dimensional in the sense that they remained close to the equatorial plane of the Saturnian system, like the Galileo mission in the Jovian system. *(NASA/JPL)*

directly to what part of Titan the spacecraft will fly over – and hence where the RADAR or other instruments will be able to observe.

Hence starting from an initial orbit (configured for delivering the Huygens probe shortly after arrival in the system), there is a choice of cranking and/or pumping. Let's say there are roughly four choices per Titan encounter (to pump up the orbit, crank up the inclination, pump down the orbit or crank down the inclination) and that on average Titan encounters occur once per month or so, then in a four-year tour of 48 encounters there would be (in principle) four raised to the 48th power of possible tours, which is about 8 with 28 zeros after it!

Another option, feasible only in certain equatorial orbits, would be to skip a fraction of an orbit and flip from inbound to outbound Titan flybys – this non-resonant transfer allows the direction of the apoapsis petal to be rotated around in the orbit plane.

After polling all the science teams for 'wish lists' of the opportunities for observations that they deemed essential in a tour, the mission designers generated about four dozen permutations of various options in 1994 and sought evaluations and feedback. JPL engineers also provided science teams with a spreadsheet tool called STOCK to enable them to devise coarse tours of their own. This allowed the various teams to identify the most that they could possibly hope to achieve – if the mission were to be skewed entirely in their favour. For example one enterprising radar engineer, focusing only on maximising opportunities for that instrument to observe Titan, managed to generate a tour with 72 close flybys. Of course, such a tour would have provided little geometric diversity on other targets, and severely limited the opportunities to visit other satellites.

The next iteration saw the generation of 18 candidate tours labelled T1 to T18. The process required the scientists to learn what was possible subject to the constraints of astrodynamics (just how much Titan's gravity could change the orbit of the spacecraft) and project operations (limiting the delta-V to 500m/s, having no more than four consecutive 16-day orbits which would lay a heavy workload on operations staff, and so on), and required the tour designers to understand the relative

importance of the different science observing opportunities. For example, after much discussion the 'shopping list' of icy satellite encounters was determined to be (in priority order) Enceladus-1, Iapetus, Enceladus-2, Dione, Hyperion, Rhea, and Enceladus-3. The moon Enceladus was of particular interest because of its suspected geological activity and apparent relationship to the E-ring.

As this effort proceeded in the years prior to and just after launch, a couple of the T-1 to T-18 set were selected for further evaluation, leading to variants T-9-1 and T-18-1 to T-18-5. It was recognised at this point that the most efficient tours would rotate the orbit-petal clockwise, meaning that as Saturn travelled around the Sun in its 29.5-year orbit, the Saturn-Sun line would rotate clockwise by ~12°/year, or 49° during the four-year prime mission.

A fortuitous alignment was discovered in the satellite positions, exploited in both the T-9-1 and T-18-5 tours, whereby an orbit just after the seventh Titan flyby would skirt both Hyperion and Dione prior to re-encountering Titan three-and-a-bit Titan days later. This was a rare non-resonant sequence where the encounter with Titan switched from the outbound leg to the inbound leg.

Broadly, the tours still in contention all made the same four major steps. First, pumping down the orbit period from the initial long arrival and probe delivery orbits. Then the apoapsis petal would be rotated to the magnetotail. The next step would rotate the petal to the Saturn day-side for imaging. And finally the inclination would be cranked as steep as possible (about 65°).

Another possibility for moving the petal was not to rotate it slowly around in the equatorial plane but to raise the inclination, then flip from inbound to outbound and then crank the inclination back down. This trick was called 'cranking over the top', and also described as a pi-transfer because the inclined flipping corresponded to a change of true anomaly (or orbital phase) of exactly 180° or pi radians. This sequence of flybys yielded 135° of petal rotation in a period of about a year, but provided many more Titan flybys and more diverse viewing geometry than the equivalent equatorial rotation.

The different science teams examined the opportunities provided by each of these tours.

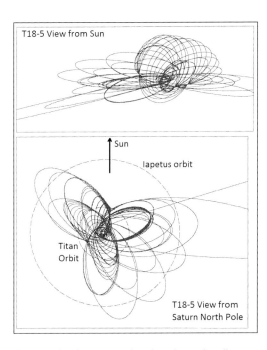

RIGHT The T-18-5 tour was the basis for that which was actually flown. The upper panel is a view from the Sun, with the long looping arrival orbit at the left. The 'cranking over the top' sequence of orbits nicely illustrates the three-dimensional nature of the tour. The lower panel is looking from Saturn's north pole (with the initial orbit at the right) and shows the distribution of local times or angles relative to the Sun of the orbit petals.
(NASA/JPL)

This required computational tools to visualise the observing geometry throughout the four years of the tour, from which different quality metrics could be determined (e.g. the length and distribution of close ground-tracks on Titan for radar coverage, and the number and geometry of ring occultations). Of course, not only were different science teams in favour of different tours overall, but even within a given team there could be conflicts (e.g. those interested in studying the satellites versus the rings). A fundamental difficulty that was never really resolved by this process, was that only the opportunities could be assessed – because Cassini couldn't do everything at once (e.g. aim its antenna at Earth for tracking while also pointing the UVIS instrument at the Sun for an occultation measurement). Just because a given tour had a beautiful geometry for one type of observation didn't mean this observation would actually be performed.

Despite this challenge, the scientific inputs guided the selection process. Other factors came into play of course, such as minimising the timing of periods of high activity such as flybys or engine burns on major holidays, and minimising the propellant required to adjust the orbit. Major activities had to be avoided at times of solar conjunction, when Saturn was behind the Sun, or even very close to it in the sky, since the Sun is a prominent source of radio noise.

The feedback from the science teams on

the relative merits of the different tour options was assimilated, mediated in part by discipline-focused working groups chaired by Cassini's Interdisciplinary Scientists, who had been chosen not to work with a specific instrument but to try to coordinate synergies between instruments.

Eventually, T-18-5 was settled upon. It had 44 Titan flybys over a four-year period (of which 17 included Earth occultations) and the full set of close encounters with the icy satellites.

It was on the basis of this tour that in the early 2000s the highly contentious process of carving up the tour into observing opportunities for each instrument began. Frustratingly, it later proved necessary to make a significant change to the first few months of the tour in order to accommodate an altered delivery scenario of the Huygens probe designed to reduce the Doppler shift during the probe relay. This revision added an extra Titan flyby early on, as well as fortuitous Iapetus observations, and managed to resynchronise with the planned tour so that most of that development could be retained.

With the four-year Prime Mission (PM) underway, the trajectory planners set to work to define an additional two years of activity in the Saturnian system which was initially referred to as the Extended Mission (XM) but later called the Equinox Mission (EM). This would aim to address science goals incompletely tackled during the PM. For example, much of the Titan coverage of the PM was in the northern hemisphere, and so the XM would introduce more flybys in the southern hemisphere. The discovery by Cassini of plumes on Enceladus was motivation to add flybys of that moon, particularly to observe its south polar region.

The XM tour would start in the highly inclined orbit in which the prime tour ended, and after spending some time in inclined orbits it would undertake a pi-transfer and then crank down the inclination. This inclination reduction from flybys T52–T62 would give a set of nice, near-parallel low-altitude flybys of Titan to observe its poorly mapped south-western quadrant, and would occur around the time of Saturn's equinox in 2009, when the Sun would be in the ring plane. Once Cassini had resumed flying in the ring plane it could make a number of further satellite flybys, including Enceladus. The EM was to include no fewer than ten new close satellite flybys, of which seven would be Enceladus.

The so-called Extended-Extended Mission (XXM) that was later officially named the Solstice Mission (SM) posed particular new challenges for the tour designers.

First, because the propellant budget was now very low, the engine burns would require to be kept to a minimum. Any flexibility in timing (e.g. shifting encounters by a few days in order to save 1–2m/s of delta-V) would be exploited to the utmost.

Second, the discovery of plumes on Enceladus now implied that there might be a habitable environment there. As a result, planetary protection considerations became significant. Specifically, by international treaty, spacefaring countries have agreed to avoid contaminating extraterrestrial environments with life from Earth, lest this threaten any traces of alien life. In some cases it is merely adequate to document the bioload (the number of culturable bacteria per square meter of surface of a spacecraft, then demonstrate by calculation that the chance of a terrestrial cell surviving that environment is 0.01% or less). In others, the planetary protection burden is more significant, requiring sterilisation of the spacecraft by a method such as dry baking. Since Cassini carried RTGs which could (in principle) melt a small habitable pocket in Enceladus's ice, the spacecraft could not be left in a state where it might one day encounter Enceladus. Given the chaotic evolution of orbits, in reality this meant Cassini could not be left in

orbit at all. So the decision was made to dispose of the spacecraft by plunging it into Saturn (as Galileo had been at Jupiter, in order to ensure that Europa could not be contaminated in this manner). Thus, the XXM design had to include a disposal scenario in 2017.

The emphasis of XXM science centred around several major objectives. Most of all, the opportunity to study seasonal variation in the Saturnian system formed a key part of the rationale. Eking out operations to 2017 would require that operations be less intensive than before, with fewer flybys per annum than in the PM. In particular, science activities needing fast paced turns using the hydrazine thrusters would be severely restricted in order to minimise propellant usage.

Even so, towards the end of the XXM the spacecraft would be running perilously close to empty. If the bipropellant tanks actually ran dry – which was a possibility because it was impracticable to measure the amount of fluid remaining in a tank in a state of weightlessness and it could only be estimated indirectly from engine firing times, tank pressures and temperatures, etc. – then some manoeuvres would require to be performed by the monopropellant hydrazine system, presuming that this had not also run out. But the hydrazine was also required for fast science turns, so its management would become a critical issue. And if something were to go wrong, such as a stuck thruster, the recovery effort could drain an

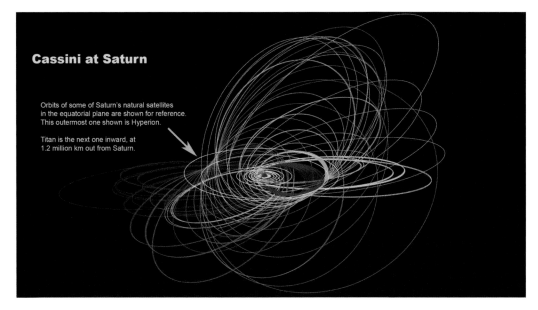

Cassini at Saturn

Orbits of some of Saturn's natural satellites in the equatorial plane are shown for reference. This outermost one shown is Hyperion.

Titan is the next one inward, at 1.2 million km out from Saturn.

LEFT The seven-year Solstice Mission of Cassini, colour coded by time. *(NASA)*

undetermined amount of hydrazine. Similarly, if the reaction wheels started to fail, it would be necessary to employ the hydrazine system instead. These concerns weighed heavy on mission planners in 2009 as the XXM was being designed because they were planning an additional seven years of operation for a spacecraft which had already been in space for 12 years and hence was out of warranty. And of course, consideration had to be given to its safe disposal in the event that the tour was curtailed early.

Around the time that the XXM orbit was being designed for Cassini, NASA selected a mission called Juno to explore Jupiter's structure and composition using gravity and magnetic field measurements, as well as microwave radiometry investigations. The Juno spacecraft would have to be on a highly elliptical orbit, making close passage to Jupiter to obtain its scientific measurements, while loitering in a high apoapsis beyond the radiation belts in order to transmit its data to Earth. The endgame of the Cassini tour in 2017 would feature similarly elliptical orbits with close periapsis passages, the last of which would penetrate Saturn's atmosphere to cause the spacecraft to burn up.

Diagnosing and fixing the Huygens receiver anomaly

On 3–4 February 2000, a communications test that was widely thought to be an unnecessary distraction was performed by sending a stream of data on the Huygens radio frequency from the 34m Deep Space Network antenna at Goldstone in California to Cassini. All of the individual components of the system had already been tested extensively on the ground prior to launch. The datastream to be transmitted was generated by a laptop computer which replicated the telemetry from the Huygens Engineering Model at the European Space Operations Centre (ESOC) in Darmstadt, Germany.

At the time, Cassini was 430 million km away, passing through the asteroid belt, and so the radio signal took 48min to reach the spacecraft. The receiver in the Probe Support Avionics (PSA) locked on to the transmission and extracted the datastream into packetised telemetry, which was then transmitted back to Earth for an end-to-end check.

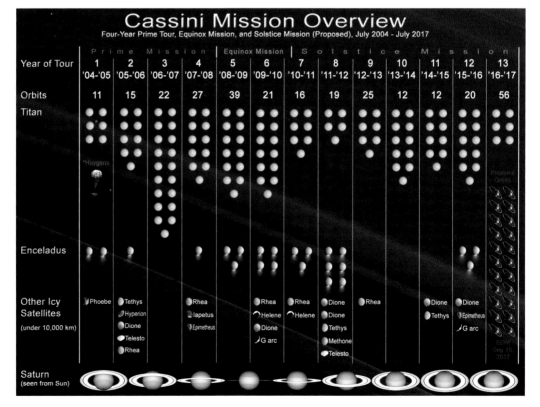

RIGHT The 'scorecard' for the Prime, Equinox and Solstice Missions, showing the Titan, Enceladus, and other close encounters, together with the view of Saturn from the inner solar system. *(NASA/JPL-Caltech)*

Unfortunately, the recovered packets were gibberish. The PSA housekeeping confirmed that the receiver had locked onto the signal, but the bit sequence was incorrect apart from a few lucid intervals. Strangely, increasing the transmission power sometimes helped, but sometimes did not. Abbreviating some tests in order to explore a hunch, the ESA engineer tried a sequence in which the simulated Doppler shift on the radio signal was removed – that worked!

The test data, supplemented by further tests on the Huygens engineering model, pointed to the digital tracking loop in the receiver. It was realised that whilst the Doppler shift had been correctly taken into account for locking onto the carrier signal, the bandwidth of the bit synchronisation part of the receiver was too narrow to allow for a change in the bit rate. The changes involved were merely a few parts per million; instead of 8,192.000 bit/s the bitstream would be coming in at about 8,192.1 bit/s. After a few seconds, the loop would be half a bit out of sync.

Unfortunately, while this could easily have been made an adjustable feature, the receiver built by Alenia Spazio of Italy had the bandwidth too small and hard-coded in logic that could not be corrected from the ground. Despite design reviews, the fact that the unit was only designed to handle 2kbps and the Doppler effect on symbol rate was not explicitly stated in the specification was not noticed until the in-flight test.

At least time was on the project's side. There were several years before the receiver would need to do its thing, allowing time to properly understand the issue and to devise and validate a solution. A redeeming feature of digital receiver design is that the system is deterministic. Armed with the design specs for the receiver, and tests using the full-up Huygens engineering model at Darmstadt, a computer model of the behaviour was quickly developed and checked against three further in-flight relay tests in the year after the Jupiter flyby.

Huygens assembled individual 126-byte packets of data from each experiment, or of probe housekeeping data, into frames according to a 'polling' table that was a list of data sources polled in sequence in 'round robin' fashion. In essence, the number of entries each experiment had in this list determined

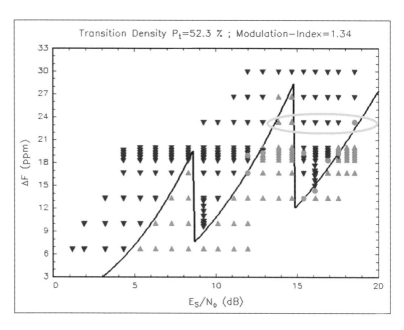

ABOVE This scatter plot of in-flight tests of signal strength (E_s/N_o) against frequency shift shows the irregular success/failure boundary (blue line) predicted by the receiver model, against the results of the in-flight relay tests (green triangles indicate success). The model successfully predicted the failure envelope – including the surprising areas ('shark's teeth') where increasing the signal strength led to loss of telemetry due to activation of the automatic gain control in the receiver. Although this exercise showed that the Huygens mission as-designed (orange ellipse) would lose most of its data, the understanding developed from this analysis paved the way for a solution. *(ESA)*

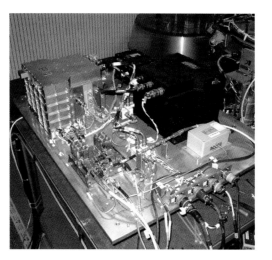

LEFT The probe support avionics of the Engineering Model, mounted on a table near that vehicle. The two large boxes were the receivers (the small black box at back-right was the RUSO). As evidenced by the extra wiring and tools, the EM equipment was intended for tests. *(Author)*

BELOW A simplified schematic of the data transmission process. The RF channel caused the symbol rate to shift out of the digital transition tracking loop bandwidth, producing to cycle slips that caused the Viterbi decoder and frame synchroniser to fail. *(ESA)*

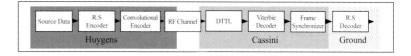

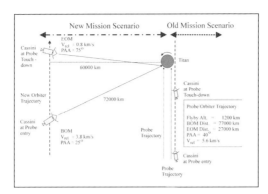

RIGHT The relay flyby geometry was a major part of the fix. Having Cassini fly past at a larger distance (60,000km instead of 1,200km) reduced the projection of Cassini's velocity onto the line of sight, lowering the Doppler shift. This also required retargeting the entry point of the probe. *(ESA)*

BELOW The redesigned tour shortened the first two orbits and inserted an additional one, with delivery occurring on the third flyby, instead of on the first. A number of manoeuvres had to be executed between Saturn orbit insertion and the probe delivery. Note that when the probe was released, Cassini was actually arcing away from Saturn prior to falling back to Titan entry three weeks later. *(NASA/JPL)*

its effective data rate. The Probe Onboard Software (POSW) assembled a transfer frame each second in each CDMU channel such that each frame contained seven packets, together with header information and Reed-Solomon (255,223) error detection/correction codes, together adding up to 8,192 bits. The bitstream had a further (R=1/2, k=7) Convolutional Code applied prior to being modulated onto the carrier using Binary Phase-Shift Keying (BPSK).

Once the receiver in the PSA aboard Cassini had locked onto the carrier signal it used a digital transition tracking loop (DTTL) to extract the datastream, performed Viterbi decoding to undo the Convolutional Code, and then detected the frame synchronisation marks to identify which bit was 'first'. These frames would then be sent to Earth, where the Reed-Solomon decoding would be performed by ESOC in order that the experiment packets could be extracted and distributed to the experimenters.

The small changes in the symbol rate presented to the receiver (about 20 parts per million) owing to the Doppler shift meant that

the tracking loop would fail. Exactly how often this would occur, and how that depended on the signal strength and other factors was diagnosed with the computer model, validated by further tests with the Engineering Model in Darmstadt and Cassini in-flight, and then an assessment was made of how well the loop would perform under different scenarios.

A major part of the 'fix' for the problem was to execute the Huygens relay with Cassini passing by Titan at a greater range than originally intended, so that the flyby velocity of 5.6km/s would be projected only obliquely onto the line of sight, thereby significantly reducing the Doppler shift. This required not only changing Cassini's trajectory, but delaying the probe delivery by several months and shifting the landing site.

The tour redesign not only needed to accommodate the new probe mission scenario, it also had to allow for a backup in case of any new late problems, and to sync-up with the already designed Cassini orbital tour as early as possible in order to minimise the amount of replanning.

The solution was to insert an additional orbit at the beginning of the tour and delay the probe delivery by several months. Instead of the first flyby (T1) in November 2004, Cassini would carry the probe until its orbit around Saturn had been pumped down by two Cassini flybys (now called TA and TB) to reduce the relative velocity for the probe relay on TC. Then the trajectory would be adjusted to match that intended for the original T3 orbit so that the remainder of the original tour T-18-5 could be performed.

The observing plans for TA and TB had therefore to be specified. The additional Titan flyby early in the tour was a welcome bonus, and the overall amount of rework was modest. Essentially by accident, it was realised that the probe release orbit would actually take Cassini (with Huygens now in free-flight) within 60,000km of Iapetus. This was a tremendous bonus since there was only one flyby (in September 2007) planned for this hard-to-reach satellite and this additional 'untargeted' flyby would allow Cassini to view the opposite hemisphere from that of the scheduled flyby.

Of course there was a price to be paid in addition to the replanning effort. The additional orbit changes, which mostly involved increasing the SOI burn by 16m/s and the Periapsis Raise

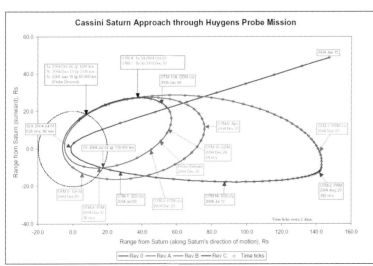

Manoeuvre by 59m/s, would cost some 87m/s of precious fuel. When the redesign was being evaluated, the fuel margin for the four-year tour was estimated at 202m/s – such a healthy margin arising in part from the delay in launching the mission because of the need to clean insulation out of the probe's interior! – so it was affordable, but it would reduce margins against future problems or for a possible extended mission.

Redesigning the orbit was actually not the only part of the fix. An additional mitigating measure was to activate the probe early, permitting it to warm up. This early switch-on, 4hr before entry rather than the 15min required to warm up the TUSO, was made possible by the healthy battery energy margins (by this time, the experience of the Galileo probe had given favourable performance results on the same batteries carried by Huygens). Hence the probe electronics would be a little warmer, shifting the frequency of the data clock by a small but beneficial amount. But this early activation in turn required the software on the probe's instruments to be updated and tested on the Engineering Model. This needed now-obsolete Ada compilers to be reactivated – and in at least one instance, a programmer to be brought out of retirement. This was not an exercise that experiment teams had budgeted for, and not every sequence could be fully optimised.

An additional 'fix' that was explored but not adopted, was that the cycle slip probability depended on the number of transitions (Pt) in the datastream; sending a continuous stream of zeroes, for example, would help the synchronisation in order that subsequent data would be retained. This could have been achieved by inserting a dummy instrument into the polling table. But because the instruments were rather efficient at encoding information in the bitstream (i.e. for Pt=50% the ones and zeroes were equally likely) such a ruse would simply have meant data being 'wasted' on the dummy sequences.

The implementation of the revised trajectory with reduced Doppler shift also required a software tweak on Cassini. By default, the PSA would look for the probe signal at a 5.6km/s Doppler shift and so if the receiver lost lock it would struggle to do so with the lower frequency in the new mission. Fortunately, a receiver mode for testing existed in which there was a near-zero Doppler shift. The

orbiter software was updated to command the Huygens receiver into this special BITE mode every 12sec while executing the relay.

Although other difficulties would ultimately confront the reception of the probe's data by Cassini, the identification and resolution of the Doppler problem was an impressive mission redesign to adapt to a problem involving hardware which was already in deep space.

Spanning thermal engineering, astrodynamics and radio systems, the recovery effort helped to engage project scientists and engineers during the seven-year cruise to Saturn and also underscored the wisdom of maintaining a fully operational Engineering Model of the probe and its payload.

Planning Cassini observations and operations

With the geometry of Cassini's path through space determined (and available as an electronic trajectory file as well as a descriptive 'Tour Atlas' document), the many science teams could begin planning their observations. To make this monumental task manageable, the tour was broken up into blocks where one science discipline or another would probably dominate.

For example, the 48hrs or so of a 'segment' (the exact time depended upon when Cassini was visible from a given DSN station) around a Titan flyby would be handled by TOST (Titan Orbiter Science Team) with corresponding working groups for Satellites, Rings, the

BELOW The green scratchy curve shows the actual flight data from Huygens, conforming essentially to the predictions (red, blue triangle curves) of the redesigned trajectory, keeping the receiver out of the pink shark's teeth. The breadth of the curve is due to signal strength changes in the radiation pattern of the antennas as the probe spun and swung beneath the parachute. The width becomes much narrower in the lower region, by which time the probe was on the surface. *(ESA)*

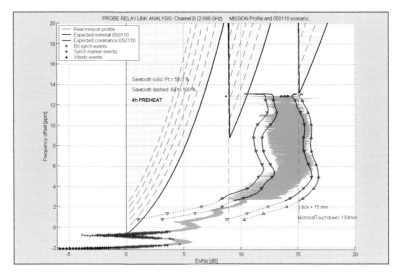

Magnetosphere and Saturn, plus a catch-all 'cross-discipline' group.

The boundaries between these chunks of time were typically defined by downlinks, when Cassini would spend around 8hr emptying its now-full data recorders to 'clear the slate' for the next set of observations.

Within these observing periods, the groups of scientists would hammer out a compromise between their conflicting demands. Of course, such a process could not handle each time block in isolation. Mapping Titan, for example, required many flybys, distributed across the surface – it might be that for example a given point on Titan could be observed on flyby T8 and T12, but if the solar illumination geometry was better on T12 then that would logically be devoted to optical mapping by VIMS and T8 devoted to radar mapping. (With these instruments viewing at right angles, it wasn't possible to use them both to observe the same target simultaneously.) So far, so good, but say T8 also had a radio occultation, an observation that was also incompatible with radar observation. So if the radar wasn't assigned the T8 flyby, it would want T12. In that case, where would VIMS be able to observe...? This is a supreme example of an over-constrained problem. Although a market-based methodology was proposed to resolve this challenge, in the end an unseemly but familiar 'smoke-filled room' approach was adopted instead.

The early part of this process, actually initiated years before Cassini arrived at Saturn, was fraught with hard bargaining because everything was at stake. Self-assured scientists, each convinced that their own parochial observations were the most important thing the spacecraft could possibly be doing, sometimes engaged in stonewalling. Occasionally deadlock had to be resolved by appeal to a group of notionally dispassionate Interdisciplinary Scientists who were not closely tied to individual instruments, to take a step back and view the bigger picture. Only once or twice did particularly contentious issues have to be resolved by intervention from the Project Scientist.

Over the years of Cassini operation, the scientific gladiators in this arena learned each other's priorities and the most efficient compromises. Working jointly in countless long teleconferences and meetings to develop the 'pointing timeline', they became a progressively more congenial and efficient team, and the science program for the Solstice Mission (when, perhaps there was less scientifically at stake, as well as the teams and their tools being honed by practice) was developed fairly smoothly.

The individual flybys were planned at first broadly by the rule of 'optical remote sensing on the inbound illuminated leg, radio bistatic observation at closest approach and then a UVIS stellar occultation on the way out', and then progressively refined down to the minute, years in advance. At any given moment the pointing sequence would be determined by one instrument team, although they would take into account the needs of others. The time at which one team handed off to another, and the spacecraft attitude at that moment, had to be negotiated.

While each team employed their own tools for visualising the operation of their instrument, a software tool created for all teams was the Pointing Design Tool. This program would ingest a sequence of high-level commands related to instrument pointing (e.g. turn the vehicle –Z axis towards a point on Titan at 15°N, 240°W, or scan from the centre of Saturn at 2°/min towards the celestial north pole) and interpret these commands using a model of the kinematics of the spacecraft. This enabled the turn times between pointings to be accurately estimated.

The PDT was managed by the Cassini project, together with dedicated Sun SPARC workstations known as Science Operations and Planning Computers (SOPC). The SOPC

BELOW The four-year Prime Mission broken down into the command sequences S01–S41, each of which lasted about 30 days and comprised a set of science discipline 'segments' which were individually planned. *(NASA/JPL)*

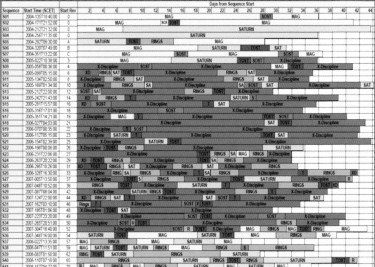

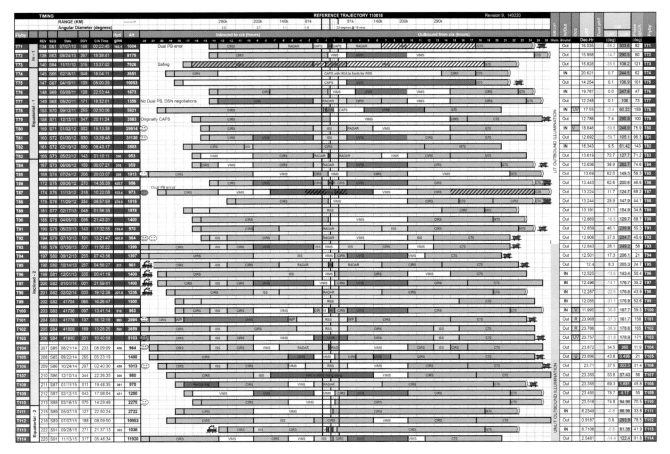

for each instrument team would be located at its operations centre – e.g. for UVIS at the University of Colorado in Boulder, or for MAG at Imperial College in London, CIRS at Goddard Space Flight Center, etc. – but because this software was considered 'Export Controlled',

ABOVE The Titan flybys of the Solstice Mission, divided into phases when an individual instrument was 'prime' and therefore determined the pointing (although other instruments were usually also able to make observations). The variable length of the TOST (Titan) segments depends on the end of adjacent segments, or the visibility of ground stations. *(NASA/JPL)*

Start Time	End Time	Prime Activity	Obs. Detail	Op Mode	TLM Mode	Comments
2012-270T00:16:00	2012-270T00:56:00	SP Turn to WP	NEG_Y to Titan, NEG_X to 115.0/-2.0	DFPW Normal		
2012-270T00:56:00	C/A-13:24:39	OD Uncertainty Dead Time				
C/A-13:24:39	-13:00	CIRS	N1 extended, TN1c	DFPW Normal	S_N_ER3	
-13:00	-09:00	CIRS	N1, TN1c	DFPW Normal	S_N_ER3	
-09:00	-02:15	UVIS	X, TN1c	RADWU	S_N_ER5a for 15 min then S_N_ER3	RADAR warmup at -09:00:00
begin custom period						
-02:15	-00:50	CIRS	TN1c	RADWU	S_N_ER3	
-00:50	-00:49	RWA to RCS Transition		RADRCS	S_N_ER3	
-00:49	-00:18	CIRS	CIRS turn to INMS attitude, TN1c	RADRCS	S_N_ER3	deadband for CIRS 0.5, 2.0, 0.5
-00:18	0	INMS	RADAR riding along, TC2a	RADRCS	S_N_ER8	
2012-270T14:35:39		CLOSEST APPROACH	NEG_X to RAM, NEG_Z to Titan (Tc2a)			High solar activity; northern latitude; near noon
0	+00:18	INMS	RADAR riding along, RADAR turn to CIRS attitude, TC2a	RADRCS	S_N_ER8	minor CIRS heating possible
+00:18	+00:35	RADAR altimetry	TN2a	RADRCS	S_N_ER8	
+00:35	+01:15	CIRS	on CIRS point at +00:35, TN1c	ORSRCS	S_N_ER3	
+01:15	+01:37	RCS to RWA Transition		DFPW Normal	S_N_ER3	
+01:37	+02:15	CIRS	TN1c	DFPW Normal	S_N_ER3	
end custom period						
+02:15	+09:00	UVIS	X, TN1c	DFPW Normal	S_N_ER3	
+09:00	+14:00	CIRS	C, TN1c	DFPW Normal	S_N_ER3	VIMS rider
+14:00	C/A+28:45:21	CIRS	A2, TN1c	DFPW Normal	S_N_ER3	ISS rider
C/A+28:45:21	2012-271T19:36:00	OD Uncertainty Dead Time		DFPW Normal		
2012-271T19:36:00	2012-271T20:16:00	SP Turn to Earth for downlink		DFPW Normal		
2012-271T20:16:00	2012-272T07:46:00	Canberra 70M		DFPW Normal	RTE_N_SPB	
2012-272T07:46:00	2012-272T09:46:00	Madrid 70M		DFPW Normal	RTE_N_SPB	Dual playback for RADAR/INMS, -00:05 to +00:05
2012-272T09:46:00	2012-272T10:26:00	SP Turn to WP	NEG_Y to Titan, NEG_X to NEP	DFPW Normal		
2012-272T10:26:00	2012-272T14:26:00	ISS	ISS mosaic at first, then sit and stare for CIRS and VIMS	DFPW Normal	S_N_ER3	
2012-272T14:26:00	2012-272T18:06:00	ISS	ISS mosaic at first, then sit and stare for CIRS and VIMS	DFPW Normal	S_N_ER3	
2012-272T18:06:00	2012-272T21:06:00	ISS	ISS mosaic	DFPW Normal	S_N_ER3	
2012-272T21:06:00	2012-272T21:46:00	SP Turn to Earth for		DFPW Normal	S_N_ER3	
2012-272T21:46:00	2012-272T23:16:00	Ybias window		DFPW Normal		
2012-272T21:16:00	2012-272T23:16:00	Canberra 70M		DFPW Normal		
2012-272T23:16:00	2012-273T06:16:00	Canberra 34M		DFPW Normal	RTE_N_SPB	

LEFT Each sequence (here Titan flyby T86) had to be laid out in detail months or years in advance, with not only the scientific observations themselves, but the turns between attitudes, switching between different spacecraft attitude control modes, and downlinks (in pink) planned to the minute. *(NASA/JPL)*

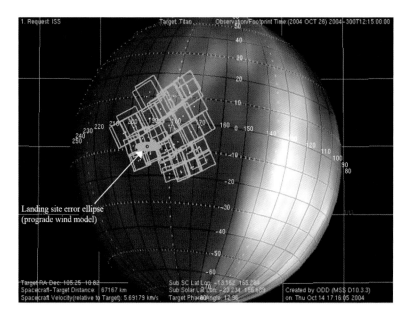

Landing site error ellipse
(prograde wind model)

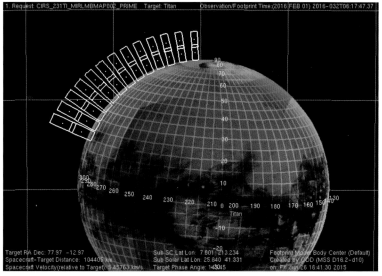

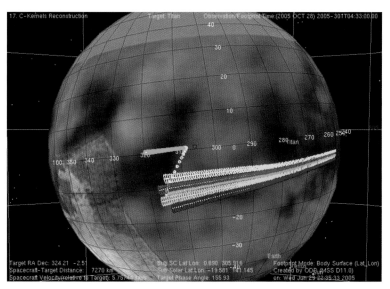

LEFT While the planning process later in the mission would benefit from Cassini's own earlier observations, the first flybys had to be planned on the basis of maps made from Voyager data or, for Titan, from the Hubble Space Telescope. This plot shows the square-ish footprints of Cassini's ISS camera during the first Titan flyby around the location where the probe would be delivered four months later. *(SSI/U. of Arizona)*

CENTRE The CIRS instrument scanned the limb of Titan in order to profile the abundance of chemical compounds with latitude and altitude in the rapidly changing northern spring polar atmosphere. The vehicle was rolled about one axis to maintain the detector array aligned along the local Titan vertical. The observation began at a low latitude (at a greater distance, hence the detector footprints were larger) and arced polewards. *(C. Nixon/NASA/GSFC)*

these computers were in locked rooms, with only designated individuals having access. These security aspects, and the need to maintain stable hardware, operating systems and software at these multiple sites over two decades, were a significant challenge.

After the pointing designs for the various instruments were concatenated together into the full observing sequence, constraint checks were applied on the integrated timeline by JPL to check for violations (such as pointing the radiators at the Sun) and estimate the hydrazine usage and/or RWA speeds.

The finite (and progressively declining) power from the RTGs prevented all instruments from being fully powered-up at any one time, so the allowable combinations of instruments for power and telemetry had to be considered. And

LEFT Radar imaging passes dragged the five radar beams across Titan's surface, looking somewhat to the side as Cassini flew right to left (here the tracks of the beams are displaced left-looking, about 10° south of the equatorial ground-track). At the end of the SAR observation, the central beam #3 was returned onto the sub-spacecraft ground-track to perform vertical altimetry (green) on the equator. *(NASA/JPL)*

RIGHT For radio science experiments such as gravity tracking or occultations, timing was crucial to ensure Cassini was well above the horizon of at least one of the Deep Space Network stations at Madrid, Spain (M), Goldstone, California (G), or Canberra, Australia (C). Here the spacecraft was handed off from Madrid to Goldstone, with much of the pass then being covered by Goldstone and Canberra (providing some insurance against weather problems, for example). The relevant elevation was not when the encounter itself happened, but when the signal reached Earth, allowing for the one-way light-time of 80min or so. *(NASA/JPL)*

CENTRE The Pointing Design Tool planning of a UVIS occultation of Enceladus's plumes. Although the tool portrayed the surface as it was then known from ISS mapping, the plume was not represented directly. The slit of the UVIS instrument was aligned so that two stars (in fact, two stars in Orion's belt) would be observed, with their light passing through the plumes on its way to the spacecraft. To the lower left is Alnitak, an O9.7Ib class star which is the brighter of the two in the ultraviolet, although not in the visible range. The star in the square box (the field of view of the High Speed Photometer) is Alnilam, a B0Ia star. *(C. Hansen)*

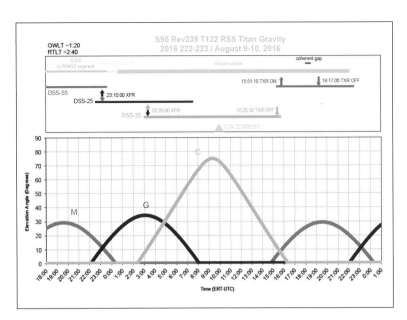

finally, because the instruments could collectively produce more data in a few hours than could be stored by the SSRs, the data volume (typically ~3Gbit/flyby) was budgeted and passed out to the different instruments according to the scientific priority of each observation.

At any one time, there would be several 'virtual teams' (often comprising the same individuals) working in parallel, for example drawing up the early design of one flyby whilst pursuing the detailed negotiations of an earlier one and the final integration and verification of another.

The sheer magnitude of this process meant

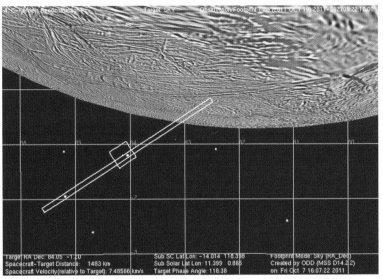

RIGHT A lawnmower pattern on a spinning planet. This radiometry of Saturn just swept the beam north-south with rather little east-west shift in inertial space, as indicated in the deep space views at top and bottom. Meanwhile the rapid rotation of Saturn meant that successive sweeps were displaced in longitude, thereby mapping the coverage shown. *(NASA/JPL)*

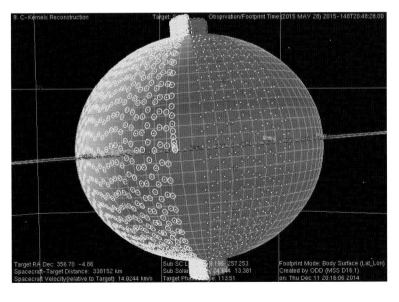

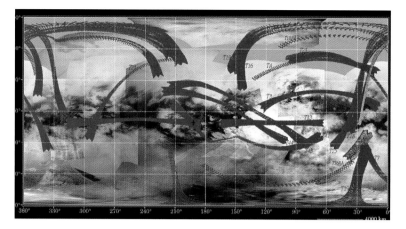

that the finite workforce could not replan Cassini observations especially nimbly. However, in the event of discoveries (such as the plumes emanating from Enceladus), individual instruments could adjust their plans within the overall schedule and data volume constraints as long as this was done several weeks in advance, in time for the new 30-day sequence to be integrated and uplinked to the spacecraft. But on a mission lasting years, there was in the end ample flexibility to adapt to instrument performance and scientific discovery.

Wear and tear on Cassini's wheels and thrusters

Unlike many spacecraft, such as communications satellites which simply maintain a fixed orientation to point their antennas towards Earth, Cassini was subject to very diverse orientation demands. At times, it would point steadily at Earth for 8hr (the interval when it was visible above a ground station) in order to send data back or to conduct a radio science experiment. At other times, Cassini had to be very agile, for example rastering back and forth like a lawnmower in the sky to build up a mosaic of images, or slewing rapidly to track an area on Titan or Enceladus as the spacecraft swept rapidly past.

Being the size of a small bus, Cassini proved to be a very stable spacecraft – the small nudges from solar radiation pressure, magnetic fields and so on took a long time to perturb the massive spacecraft's attitude. This made it an excellent platform for taking long exposure images. On the other hand, its inertia made it slow to respond to control inputs.

There were two main systems on Cassini for controlling its attitude – thrusters and the Reaction Wheel Assemblies. Reaction wheels are like flywheels in that a motor spins the wheel faster or slower and the torque to achieve this reacts upon the vehicle, prompting it to turn in the opposite direction.

Reaction wheels were the preferred means of controlling the orientation of the spacecraft because they required only electrical power. Also, because the speed of the motor could be precisely controlled they could maintain very accurate pointing. But the motor speed had an ultimate limit of ~2,020rpm and the resultant slew rate of the vehicle when operating 'on wheels' was generally limited to 1.6mrad/s. It was possible to rotate at ~3mrad/s around the High-Gain Antenna boresight or Z axis because the mass of the spacecraft was arranged more compactly about this axis, so the moment of inertia was lower.

Of course, the mass properties of the spacecraft evolved as propellant was consumed and, indeed, when the Huygens probe was released. In around 2007, the X, Y and Z moments of inertia were 7,110, 5,900 and 3,670kgm^2 respectively.

For faster turns, the vehicle used a Reaction Control System (RCS) of thrusters that were located away from the centre of mass in order to provide a torque. Applying a constant thrust on appropriate thrusters or thruster pairs would cause the spacecraft to accelerate its rotation about the related axes. In principle, very high spin rates could be achieved, but in practice even 'fast' slews were limited to 8.7mrad/s (about 0.5°/s) because thrust requires the expenditure of the finite budget of monopropellant hydrazine, and the faster the turn rate, the more thrust is needed to slow down again.

Also, because some external perturbation torques accumulate over time, even when nominally in a fixed attitude the wheel speeds would have to progressively increase in order to compensate. Consequently the RCS had to periodically 'dump' angular momentum (the RWAs only exchanged momentum between the spacecraft and the wheels). The thrusters were also needed in situations where large disturbance torques were caused by gas drag, when the slightly asymmetric spacecraft flew through Titan's upper atmosphere or the plumes of Enceladus. In these cases the thrust was pulsed on-and-off to counter the drag, with the fraction of 'on' time or 'duty cycle' increasing as the perturbation grew at the lowest altitude.

To put these torques in perspective, the tiny radiation pressure of sunlight reflecting off the boom of the magnetometer at Saturn's distance from the Sun caused a torque of about 2μNm, whereas the magnetic moment of the spacecraft interacting with Saturn's magnetic field like the needle of a compass produced typical torques a hundred times smaller than that. During a close approach (~1,000km altitude) at Titan, the aerodynamic drag of the tenuous upper atmosphere could cause a torque exceeding 0.5Nm (roughly that due to the weight of an apple held at arm's length on Earth).

It was possible to maintain a fixed pointing using thrusters, but it was less precise than using the wheels because the thrusters could only be 'on' or 'off'. A deadband was defined around the desired attitude, and when the spacecraft drifted outside this box a short pulse of thrust (an 'impulse bit') was applied to send the spacecraft back towards the centre of the box. If the pulse was strong, it would overshoot in the other direction and another opposite thrust pulse would be applied, and so on. As a result the spacecraft would slowly oscillate within a two-sided limit cycle around the deadband.

Making the impulse bit as small as possible provided more precise control and required fewer thruster firings. In practice this form of control was limited by the response of the valves, and the time for the decomposition reaction in the combustion chamber to be stable – pulses shorter than about 7ms were unreliable. Furthermore, a very small impulse bit meant lots of thruster firings, and the thruster valves were rated for only 273,000 cycles.

In fact, clever software uploaded early in Cassini's cruise to Saturn allowed it to adapt the impulse bit in the presence of a disturbing torque (e.g. closer to the Sun, the solar radiation pressure torque was much higher) so that the spacecraft would not quite overshoot the deadband, and so it would operate in a more efficient one-sided limit cycle which minimised the number of thruster firings.

Normal control of a spacecraft on reaction wheels used three wheels at right angles, one to control each Cartesian axis. (In principle it is possible to achieve limited operation using only two wheels, involving sequential rotations to point at a desired direction in space.) Because moving parts are more likely to fail in space than solid state electronics, and because Cassini's high-value mission was to last many years, it was equipped with four reaction wheels. RWA-1 to RWA-3 were installed orthogonal to each other and the axis of RWA-4 could be rotated to enable it to substitute for any of the others – making the system tolerant to the loss of a wheel.

Attitude control was performed on thrusters for the first few years after launch, with RWA operations beginning in the lead-up to the flyby of Jupiter in December 2000. It was not initially known how stable the spacecraft would be on wheels, so imaging was designed with relatively short exposures (<5sec) to minimise the possibility of motion blur. This was not a limiting concern for observing the bright planet and its satellites, but precluded long exposures that would maximise the sensitivity to light scattered by Jupiter's narrow, tenuous rings. In the event, the spacecraft was rock-solid, with these 5sec exposures typically being blurred by less than a pixel.

However, observations of Jupiter had to be suspended when one of the reaction wheels started to show signs of drag. The nine-day hiatus in operations led to the loss of some science observations. Although there were many years to go before the prime mission at Saturn would begin, this episode focused operators' attention on the wheels.

Normally the lubricant (Windsor Lube L245X, a light ester oil) in a wheel would be evenly distributed around the R10 ball bearings, but if the wheel speed happened to be very low the lubricant could collect in some places, leaving the bearings elsewhere dry, with the resultant metal-to-metal contact leading to increased drag and heating. This drag triggered an error.

Another complication was that the ball bearings were retained by a stainless steel strip or 'cage', and that at certain speeds a resonance could similarly deprive part of the bearing of lubricant. This effect, known as 'cage instability', causes vibration, wear, and possibly heating capable of degrading the oil. This could compromise the lifetime of the bearing.

RIGHT Inside each of Cassini's RWAs was the wheel itself, with the bearing around the centre. The way the bearing cage rotated, and the redistribution of lubricant around the race at different speeds, was critical to low-friction and reliable long-term operation of these systems. (NASA/JPL-Caltech)

BELOW The drag torque on the RWA-1 wheel as a function of spin rate and time. As one would expect, the drag increased with speed in either direction. But as the mission progressed, the drag on the ageing wheel at low-rpm increased. The operational procedures were revised to avoid these low-speed situations. (NASA/JPL)

This instability was first picked up in RWA-3 in October 2002 during the cruise between Jupiter and Saturn. Intermittent recurrences with increases in drag and rough operation led to the decision to take this wheel off-line in July 2003 and substitute RWA-4 instead.

A special piece of software called RWA Bias Optimisation Tool (RBOT) was developed to predict the speed of the wheels, then the manoeuvres and science observations that would lead to low-rpm states (those below 300rpm) were redesigned in order to bias the wheels at a faster speed. On the other hand, operations at close to the maximum were avoided to prevent accumulating too many revolutions on the wheels. RBOT optimised the trade-off between those factors.

The drag of each RWA was monitored regularly by spinning its wheel up to ~900rpm and allowing it to 'coast' whilst measuring the rate at which it was slowed down by friction. Of course, for such tests either the attitude of the spacecraft itself had to be able

to vary or the thrusters used to compensate the wheel spin changes. Typically 10–20g of hydrazine propellant was used on each test, executed every 90 days or so. The drag could also be estimated from the control system during regular operations, and this option was preferred during the later years of the mission because it did not expend hydrazine. Cage instability was observed on RWA-2 briefly. In fact, the vibration from the wheels could sometimes be detected as noise in data from the CIRS instrument – in effect the CIRS interferometer was acting as a seismometer!

In sum, it should come as little surprise that Cassini's wheels, parts which were moving continuously for nearly two decades in space, were precious assets in need of meticulous care. The detection of problems during the quiet cruise allowed the implementation of procedures to manage the wheels as a consumable resource. Not only was the total number of revolutions budgeted at 4 billion per wheel (these were being consumed at about 1 million revs/day, and a couple of the wheels have now exceeded 5 billion revs) but the total time spent at low-rpm was limited to a maximum of 12,400hr (in 2005–06, the wheels spent about 2.5hr/day in this regime, but the introduction of RBOT and disciplined management of the wheel speeds reduced this to an average of 21min/day in 2010–13). Like other Cassini systems, the number of power-cycles applied to the RWAs was minimised, and this averaged about 30 times per year (e.g. when main engine burns were performed). This careful operation was rewarded with all-but-flawless operation over nearly two decades, and well past the warranty.

The thrusters were also managed carefully. The obvious limitation was on the total amount of propellant available. Another limitation was the total number of thruster firings, each of which require actuation of the propellant valve.

Hydrazine monopropellant thrusters squirt hydrazine liquid, usually fed through a solenoid valve from a tank pressurised by helium, onto a catalyst bed that causes the exothermic decomposition of hydrazine into ammonia, nitrogen and hydrogen. These hot gases expand through a nozzle to produce a jet whose momentum flux corresponds to the thrust. Like any other chemical reaction, the

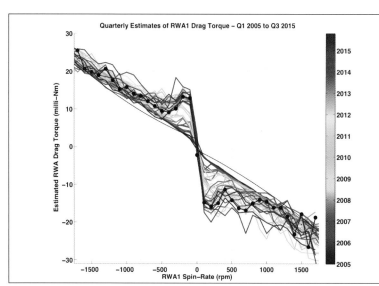

Quarterly Estimates of RWA1 Drag Torque – Q1 2005 to Q3 2015

decomposition depends on temperature. If the catalyst bed were initially cold, the thermal shock would degrade the catalyst and the thrust would start slowly and unevenly. This would yield unpredictable performance, especially for the short firings needed for a small impulse bit. Thus the catalysts are kept warm (~180°C) by electrical 'cat bed heaters' (a term that causes enduring mirth among feline-loving Cassini scientists) to ensure fast, even operation.

The thrust depends in part on the pressure in the combustion chamber, which must be less than that in the tank in order for the propellant to flow into the chamber. In some propulsion systems, the tank pressure is kept constant by bleeding in gas through a regulator to keep pace with the consumption of propellant. An approach which avoids the reliability issues of operating a pressure regulator is to operate in 'blowdown' mode. Here a fixed mass of helium sits in the headspace ('ullage') above the propellant. Its pressure falls as the propellant is used up and the helium volume expands.

Cassini used something of a hybrid between these approaches. It used blowdown mode, but because the thrusters nominally rated at 1N put out about 0.97N at launch and by the beginning of 2005 had fallen to about 0.7N, a one-time pressure 'boost' was applied in 2006 by opening a pyrotechnic valve on a helium recharge tank to restore the hydrazine pressure, which had declined to 17bar through use, back up to the launch value of about 26bar. This eliminated concerns that unacceptably low pressure would impair the performance of the thrusters later in the mission.

But eventually other issues began to appear. In particular, during an Orbital Trim Manoeuvre in October 2008 thrusters Z3A and Z4A underperformed significantly – their thrust dropping noticeably below the declining trend due to the steady pressure drop (the Z-thrusters were used more because they performed the small OTM burns as well as attitude control). The underburn imposed a 5.7m/s penalty spread over subsequent manoeuvres to correct the trajectory. Moreover, the A-side Z thrusters were equipped with pressure transducers, and these showed the operation of those thrusters to be 'rough' with large pressure fluctuations indicating poor catalyst performance. This led to

the decision in 2009 to switch from the A-side thrusters to the back-up B-side (an operation lasting several days, to monitor the attitude closely to detect thruster leaks and so on). After the switch, the B-side performance has been watched carefully, and new software tools were developed to pick out signs of thruster degradation.

The monopropellant hydrazine was held in a spherical tank with a diameter of 28in. To ensure bubble-free flow from the tank into the thrusters, the liquid had to be maintained around the exit pipe. This was achieved on Cassini's RCS by using an elastic membrane to hold the liquid in position in the absence of gravity. The diaphragm was composed of a specially formulated elastomer (AF-E-332), a polybutadiene rubber with Aerosil R-972 silane-treated silica stiffening agent to resist the corrosive propellant. The tank itself was created from sections of 6AL-4V titanium alloy, welded using special TIG (tungsten inert gas) procedures so that the welding process did not heat the elastomer above its curing temperature.

After many years in space, it seems a tiny amount of silica had leached out from the membrane and been progressively deposited on the cat beds, degrading their performance and leading to irregular pressure in the chamber and consequent variability in thrust. Like a form of arthritis, this would degrade the B-side thrusters as Cassini moved into its twilight years, but hopefully the spacecraft will be able to achieve its grand finale as planned.

BELOW RWA-4 **was mounted on a mechanism that allowed it to be rotated, enabling it to substitute for any of the fixed RWAs that failed.** *(NASA/JPL-Caltech)*

Chapter Six

Saturn arrival

Originally, the first main events in the Cassini mission at Saturn would be to insert into orbit and deliver Huygens on the first close Titan flyby. But an opportunity was found to visit Phoebe 'on the way in', and after the mission redesign there would be two close Titan flybys as well as more distant views, and a bonus Iapetus encounter. Cassini's first months in orbit would be busy.

OPPOSITE An artist's impression of Saturn orbit insertion. Only one of the two main engines fired, the other standing by in case of a problem. This 90min burn made Cassini into a satellite of the ringed planet. *(NASA/JPL)*

Phoebe encounter

As Cassini approached the Saturnian system in 2004, anticipation grew. In fact the first whispers of the target were picked up by RPWS. Bursts of so-called Saturn Kilometric Radiation, the radio emissions from lightning discharges in the planet's atmosphere, showed up in spectrograms obtained from a range of 2.5 AU. Later, long-distance observations by ISS showed the planet in colourful megapixel grandeur, and in April 2004 started to resolve a level of detail of Titan comparable to that achieved by the Hubble Space Telescope (i.e. showing it 10–20 pixels across). Over the following three months the ISS NAC pixel scale improved from 200km to ~35km and teased out new detail.

But the first dramatic science was the small outer moon, Phoebe. The fact that it orbited Saturn in the 'wrong' direction indicated it wasn't formed in the swirling disk from which most of the moons formed, but was a captured asteroid, or comet, or something in between (a population of objects, the 'Centaurs', orbit the Sun in that region). It was an appealing appetiser for Cassini's feast in the Saturnian system.

Phoebe's distance from Saturn is so large that there would be no prospect of coming anywhere close to it once Cassini was in orbit around the planet, but by timing the interplanetary trajectory just right, Cassini could fly right past Phoebe on its way in.

In fact the flyby could be made arbitrarily close, which some measurements might prefer. But a really close flyby would make optical instrument pointing vulnerable to ephemeris errors – the direction from Cassini to Phoebe was dependent on the small difference between estimates of the positions of the two objects as measured from Earth. That, and concern about whether Phoebe was surrounded by clouds of dust, led to a compromise flyby altitude of 2,000km. Post-encounter navigation data indicated the flyby had been at 2,071km (a tad high) and about 0.2sec early. Having travelled so far through interplanetary space, this wasn't bad!

The observations were spectacular. The images showed a surface pockmarked with impact craters; this was not in itself a surprise, but the craters exposed layers of different material in the subsurface. VIMS spectral data showed the distributions of different materials – water ice, iron (a common component of minerals) and carbon dioxide frost. The presence of carbon dioxide strongly implied that Phoebe didn't originate in the asteroid belt, but in a much colder region of the solar system such as the Kuiper Belt (a population of objects which includes Pluto as one of its largest members).

Getting spectroscopic data about Phoebe was also of interest in solving the mystery of Iapetus, whose yin-yang surface had been speculated to be somehow caused by infalling dust from Phoebe.

Three weeks after Phoebe was the moment of truth – when Cassini would live up to its classification as a Saturn orbiter.

Orbit insertion and the rings

Saturn orbit insertion was one of two 'critical sequences' for Cassini (the other being the Huygens probe release and relay) where particularly careful scrutiny was applied to operations. After the loss of the Mars Polar Lander in 1999 without any communication, NASA ruled that critical events had to be able to be monitored in real time.

Accordingly, Cassini switched to its Low-Gain Antenna 1hr 45min before the burn was to start, so that a radio carrier signal (without science data) could be transmitted throughout the burn to enable progress to be inferred from the change in Doppler shift, and 20min later the spacecraft turned its Hgh-Gain Antenna into the ram

BELOW VIMS spectral data with ~4km resolution of Phoebe taken from a range of about 16,000km (compared with an ISS image at the upper left), showing the distribution of carbon dioxide (orange), water ice (blue), ferric iron (red), and an unidentified material (green). Water ice was associated with the brighter regions, and the other materials were more abundant in the darker regions. *(NASA/JPL/U. of Arizona)*

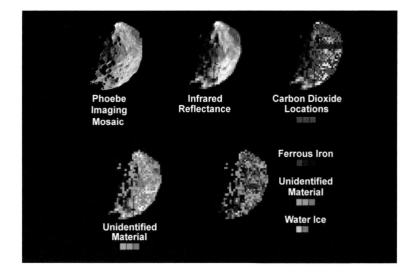

Phoebe Imaging Mosaic

Infrared Reflectance

Carbon Dioxide Locations

Unidentified Material

Ferrous Iron

Unidentified Material

Water Ice

direction to shield the instruments from possible dust impacts during the ring-plane crossing at SOI-25min. Then 10min later, it adopted the correct orientation for the motor burn to be performed by one of its 500N main engines.

The latch valves were opened just 1min before the burn began. They had remained closed during cruise because of the leaking pressure regulator – in the original plan they were to be opened for tests some 30 days before SOI. Some four years of planning and simulations had gone into designing the orbit insertion sequence in this off-nominal condition in order to make sure that the tank pressure would be in the range 195–324psi at both the start and the end of the burn, sufficient for smooth fuel flow, but not too high to breach the tanks. Not only did such analyses need to model the expected behaviour of the burn, but also ensure the onboard fault protection logic would be able to manage the tank pressures in the event of the spacecraft aborting the burn and going into 'safe mode'.

While spacecraft motor firings are often approximated in orbital designs as 'impulsive' (i.e. infinite thrust applied for zero time), the tremendous velocity change (609m/s) needed to brake Cassini into orbit from its fast interplanetary trajectory required a motor to fire over some 96.6min, while the spacecraft swept a wide arc past the planet. This long arc meant that a burn in a fixed orientation (a typical approach, with spacecraft sometimes being spun for stability during the manoeuvre) would be inefficient, and therefore Cassini had to be 'steered' to hold the thrust vector along the velocity vector as that veered through 46.4°. This steering was accomplished (as on many launch vehicles) by thrust vector control – i.e. nudging the main engine thrust. In Cassini's case this was done with an actuator that moved the rocket motor nozzle by a small amount to slew the spacecraft around. In fact, some subtle limit-cycling during the burn was noticed with a period of about 30sec, which was later diagnosed to be an effect of backlash in the motor actuator – as distinct from other small wobbles in the burn due to propellant slosh in the tanks (6sec and 16sec) and flexing of the magnetometer boom.

Controllers at JPL and anxious scientists watched the Doppler trace recorded at the DSN head off on a new trend as the burn induced a

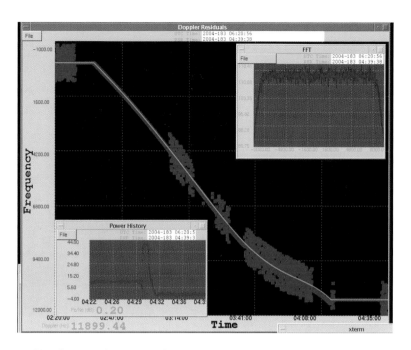

ABOVE A comparison of the Cassini radio frequency received at the DSN (red points) with the predicted history for a nominal burn (thin yellow line). The broad sections of the trace are where lock was lost, due to mispointing or to scattering by ring particles. *(S. Asmar/JPL)*

rate of change of the received signal frequency of 1–3Hz/s.

Tension was heightened by occasional disappearance of the signal, as it was blocked first by the F-ring, then the much wider A-ring and B-ring, but when the signal reappeared it was still on the required line.

After 78min of firing, Cassini had slowed to the point where it was just bound by Saturn's gravity, but the engine had more work to do. The burn was designed, unusually, to manage the spacecraft's kinetic energy onboard, rather than simply timing the engine firing. An accelerometer onboard recorded the progressive velocity change until after 1hr 36min of firing, the desired orbit was attained. The main engine cover was then closed and the spacecraft briefly transmitted engineering data reporting its status.

Cassini then turned for an unprecedented view of the rings. At this point the INMS cover was released. Science pointing was again interrupted in order to face the High-Gain Antenna forward in preparation for the second ring-plane crossing. Double playback of SOI science and engineering data took 19.5hr, with

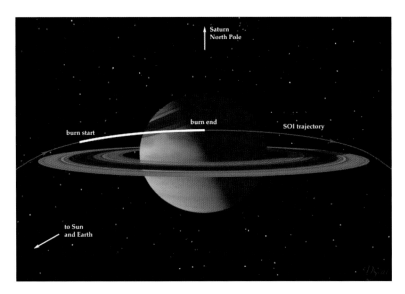

ABOVE The geometry of the Saturn orbit insertion. A portion of the burn occurred when the line of sight to Earth was blocked by the rings. The ring-plane crossings took place outside the main rings in order to minimise any impact hazards. *(D. Seal/NASA/JPL)*

BELOW The predicted scene and image acquisition parameters for an image of the Encke gap in the ring system which was to be taken shortly after orbit insertion. In the image identifier N1467351261, 'N' denoted the Narrow-Angle Camera. The 1x1 indicated that the image was at full resolution, rather than being binned 2x2. Lossless compression was applied, and BOTSIM ('both simultaneously') meant a wide-angle view was also acquired. The planning document also specified geometric information. The emission angle ~90° meant the image was taken looking almost directly down on the rings. Of course, although the simulation accurately predicted the scene geometry, it couldn't anticipate the amazing detail of wave structures in the actual image on the right. *(Planning image: Vance Hammaerle/JPL; Cassini image NASA/JPL/Space Science Institute)*

the receiving station in Madrid, Spain, handing over to Goldstone. The first images arrived about 10hr after SOI began.

The very rapidly changing geometry as Cassini swept across and over the rings required the exposure times to be very short (just 5ms), so the raw images were rather dark. However, a low signal-to-noise didn't matter for the breathtaking detail which showed up in black and white. The highest-resolution ring images showed a grainy appearance – not of individual ring particles but of gravitationally bound clusters of particles. These displayed a preferred orientation which resulted in their being nicknamed 'propellers'.

Spiral density waves and other structures – some of which had been imaged crudely by the cameras of the Voyagers, and many only identified in one dimension in Voyager occultation data – were magnificent in the Cassini pictures. Indeed, some images bore impressive resemblance to simulated images created from the Voyager profiles, but of course these downloads were real.

BELOW Despite the short exposure (yielding only modest signal-to-noise, giving rise to horizontal banding artefacts), this ISS NAC image of the unlit side of Saturn's rings after orbit insertion shows tremendous detail in the Cassini Division (named after the astronomer who first observed it). *(NASA/JPL/Space Science Institute)*

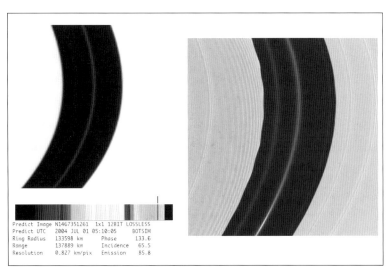

```
Predict Image N1467351261  1x1 12BIT LOSSLESS
Predict UTC   2004 JUL 01 05:10:05      BOTSIM
Ring Radius   133598 km        Phase      133.6
Range         137889 km        Incidence   65.5
Resolution    0.827 km/pix     Emission    85.8
```

The rich variety of radio signals at Saturn is seen in this RPWS spectrum obtained when very close to the planet for the orbit insertion burn: whistlers, various plasma phenomena (upper hybrid resonance, electron cyclotron), Saturn Kilometric Radiation, dust impacting on the antenna, and emissions related to the motor firing. *(W. Kurth)*

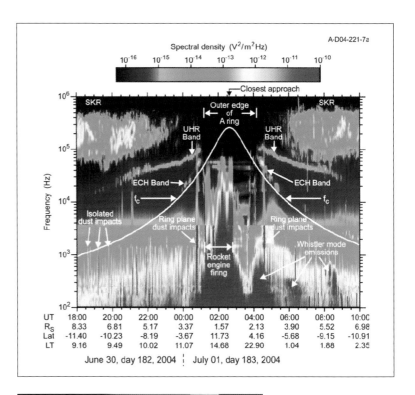

T0 – the Zeroth look at Titan

As Cassini sped outwards from Saturn to start its first orbit around the planet, it arced to the south of the equatorial plane in which most of the moons travel and on 2 July got a long-range (339,000km) view of the south pole of Titan. Although Titan observations in this phase had not initially been considered, the opportunity had to be book-kept somehow. It was hence referred to as T0, preceding T1 (or later, TA).

Only the optical remote sensing instruments would be able to make useful observations. Although some VIMS and CIRS data were taken, it was ISS that stole the show. The relatively sedate 'flyby' produced a sequence of images with a best resolution of 2km that showed a system of convecting clouds around the south pole, not only moving in the wind but also evolving in shape over the few hours between pictures. On the surface of the enigmatic moon narrow linear features were speculated to be tectonic in origin, or channels.

TA – first contact

The first 'true' Titan encounter – TA in October 2004 – was an opportunity to exercise most of Cassini's instruments. The media besieged JPL, where scientists had gathered for first contact. This would be the time to learn whether Titan's haze was too scattering for ISS to pick out small surface details; or to find that the atmosphere reached such a high altitude that later flybys might need to be replanned in order to avoid atmospheric drag; or to find that Titan crackled with lightning; or that the radar was dazzled with retroreflective scatterers (as one prominent Stanford scientist had warned in the early

This south polar mosaic from the T0 flyby shows complex but inscrutable bright and dark patterns on Titan visible through the 940nm CB3 filter on ISS. Due to the effect of looking through a longer optical slant path through the haze the image becomes progressively blurred away from the middle. The cluster of bright spots at the lower centre of the image are convective clouds around the south pole. *(NASA/JPL/Space Science Institute)*

ABOVE Scientists got their first close look at Titan on orbit TA late in the evening of 26 October, on laptops in borrowed cubicles at JPL. *(Author)*

BELOW The improved transparency of Titan's atmosphere in the near-infrared is evident in this comparison of VIMS images at 1μm (left) and 2μm (right). The 64x64 pixel image cube has a resolution of ~2km/pixel. The data were taken with an exposure time of 80ms. *(NASA/JPL/U. of Arizona)*

BOTTOM An ISS NAC view of the planned Huygens landing site (left, 0.8km/pixel, 400km across and delineated by the black box in the WAC image on the right). The surface has bright and dark markings displaying a streamlined pattern consistent with motion of a fluid such as the atmosphere. North is at about two o'clock. *(NASA/JPL/Space Science Institute)*

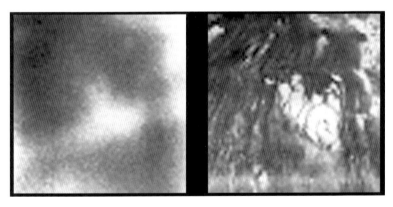

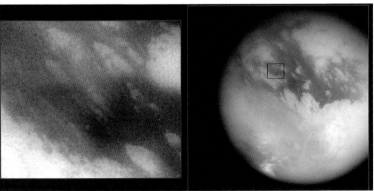

1990s). None of these things was true.

VIMS made a close-up observation (~100km wide) of a patch of Titan's surface that turned out to have a small bright feature in a darker background. There was nothing especially interesting about this spot in advance – after all, there was only the Hubble Space Telescope and a modicum of ground-based data to guide targeting of the observation when it was planned – but at 8.5°N, 143.5°W this spot was where the Huygens probe had been aimed before the probe mission had to be redesigned to accommodate the flaw in the relay receiver. Lacking any better data, the observation remained at those coordinates. The VIMS data confirmed ground-based indications that the atmosphere was much clearer at wavelengths of 2μm and 5μm than at the 1μm observed by ISS. The bright coiled feature was speculated at the time to be a cryovolcano, although subsequent evidence suggests this is unlikely.

The RADAR imaging data was somehow more of a teaser than a grand revelation. The most immediate result was that it showed lots of detail on what was clearly a complex surface which was subjected to a variety of processes. (In retrospect, the area imaged seems to have been one of Titan's most inscrutable regions.) A few features were attributed (perhaps as much for lack of a better explanation as persuasive morphology) to cryovolcanism. Several bright striated triangles connected to windy channels hinted at past rivers possessing alluvial fans. Strikingly, there were no impact craters. Surprisingly, the dielectric properties indicated by the radiometry suggested Titan was completely covered (at least at large scales) with organic compounds rather than the water ice that had been expected to be dominant.

Some of the most striking results from TA were those from the INMS. Whereas a spacecraft sampling Earth's atmosphere at Cassini's Titan altitude would see just a handful of species (molecular and atomic oxygen, nitrogen, a little hydrogen and nitrous oxide), Titan's atmosphere yielded a forest of signals at almost every mass covered by the instrument (up to 100 Daltons). Not only was there nitrogen, methane and ethane, but dozens of other compounds including benzene (mass 72). The chemistry of Titan's upper atmosphere was much richer than had been anticipated, in part because the models were limited in their ability to follow all the permutations for higher mass molecules, but also (it would appear) because ion chemistry as well as neutral chemistry was important.

Although the RPWS, MAG and other instruments were all operating, and their observations eventually contributed to the bigger picture, they detected neither a significant magnetic field intrinsic to Titan nor any radio emissions from lightning (Voyager had already placed limits on both these phenomena).

More quotidian results also emerged. The Doppler tracking of Cassini by the DSN 70m stations could measure the line-of-sight velocity to better than 0.1mm/s in a 60sec observation, and two-way ranging measurements could measure the distance from Earth to about 1m. Precision pointing, and indeed the delivery of the Huygens probe, would rely on knowing Cassini's position relative to Titan, not to Earth. TA had been 25.9km low (not a problem, given that a prudent 1,200km was targeted) and 4.2sec early, but much of this error was due to the imperfectly known position of Titan. Data obtained from TA and future passes would progressively eliminate these uncertainties and thereby improve targeting (e.g. the positional uncertainty of Titan was about 35km at TA but already down to 4km at TB).

The position of Cassini, and the orbits of the various moons, would also be refined by 'OPNAV', based upon astrometric images taken by Cassini's ISS to observe the position of bodies relative to one another, or to stars in the background. For many well-behaved

ABOVE The first eagerly awaited run of the SAR processor (after all, the entire mission was once called 'Titan Radar Mapper') showed detail but ugly stripes due to gain variations across the five beams, in part owing to ephemeris errors. However, once the correct altitude was 'dialled in', the image quality improved dramatically. The image spans about 200km. *(B. Stiles/NASA/JPL)*

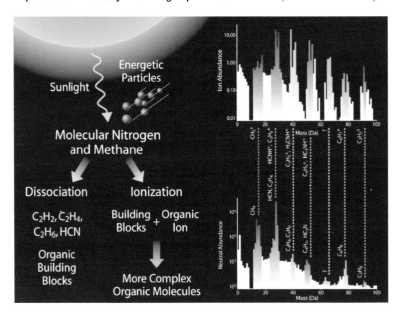

CENTRE The mass spectra of the neutral molecules and ions in Titan's upper atmosphere as measured by INMS was far richer than had been expected, with not just methane and ethane but abundant benzene and evidence from the CAPS instrument that many large complex compounds beyond the INMS mass range of 100 might be present. *(NASA/JPL/Southwest Research Institute)*

RIGHT A trim manoeuvre narrowed the ellipse of uncertainty for where Cassini would fly past Titan. The reconstructed flyby position (with Titan's limb in the lower left in this graphic) is in the plane orthogonal to the flyby velocity. *(NASA/JPL)*

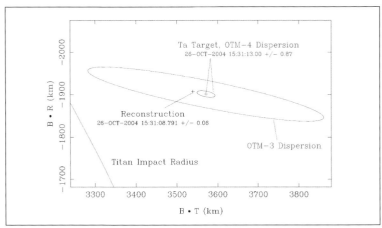

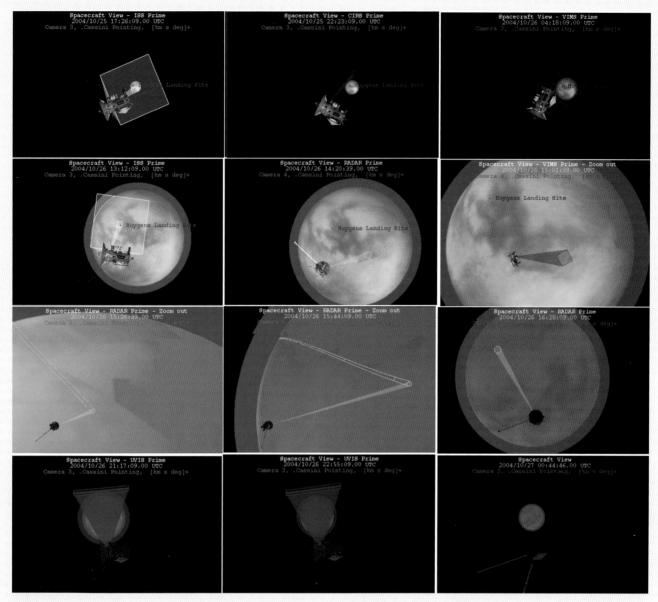

Snapshots from an outreach movie which illustrates Cassini's changing attitude for different science observations during a Titan flyby, in this case the first, TA. Initially (top left), about 20hr out, the moon already fills the field of view of the NAC (small white box) but only about one-quarter of the WAC. Next the red CIRS fields of view are scanned for atmospheric measurements (in particular to help to estimate Titan's winds in preparation for Huygens). The red box in the next frame (top right) shows the VIMS field of view for mapping the chemical composition of the surface. Although the ORS instruments are nearly boresighted, these had different mosaicking and exposure time requirements, so the observations had to be planned sequentially. Then (upper middle left) the ISS images the anti-Saturn hemisphere of Titan in some detail (white box), paying particular attention to the intended Huygens landing site (the prominent dark areas in this early ISS mosaic is Shangri-La). At this point, about 4hr out, the spacecraft is slewed 90° to point the HGA at the moon for radar radiometry and scatterometry mapping (green beam). Since the inbound part of the flyby was well illuminated, the vehicle is turned back to perform high-resolution VIMS mapping about 45min from closest approach (in this case of what became nicknamed 'The Snail' or officially Tortola Facula, which happened to be the original target of the Huygens probe). Then, the hard-working spacecraft is slewed to aim the radar antenna sideways. As the spacecraft sweeps past Titan (lower middle left, and centre) the beam footprint is dragged across Titan's surface to perform SAR imaging. Meanwhile, the INMS instrument, facing in the ram direction, measured the atmosphere in-situ at an altitude of 1,500km. As Cassini starts to recede over the night-side, radar altimetry and scatterometry are continued. The final observations (bottom left, and centre) use the slit-like field of view of the UVIS instrument to observe the upper atmosphere. Finally (bottom right) the vehicle slews to point the HGA at Earth in order to beam its hard-won data to the DSN station. (*Author montage from JPL video*)

LEFT A false-colour ultraviolet (UV3, 336nm) image of the unlit limb of Titan near 10°S at 0.7km/pixel showing many layers illuminated by light scattering through Titan's atmosphere. *(NASA/JPL/Space Science Institute)*

targets, these observations made a powerful and straightforward improvement, but Titan's dynamic haze layer and Hyperion's unusual shape and chaotic rotation made OPNAV for those bodies a little more challenging.

TB and Iapetus

The second Titan flyby, on 13 December 2004, had a very similar geometry to the first. This had to be sequenced in some haste because it had not been in the original mission plan.

VIMS observations of some mid-latitude clouds allowed the heights of the cloudtops to be measured from the shape of their spectrum. The clouds climbed and evolved on timescales of 30–60min, some dissipating as the spacecraft watched. They seemed to top out at about 44km, which was the expected maximum altitude for convective clouds, i.e. the tropopause. In all likelihood, it was raining.

Something that was not observed in the ISS images from TA and TB was a specular reflection. If there were liquids present on Titan's surface, they might be very flat and smooth and give a mirror-like glint reflecting the Sun. Dazzling in clear sky, this effect would be detectable even through Titan's haze. For the first two encounters the 'specular points' were near the edge of Xanadu, and not far from where the Arecibo radio telescope (in a record-breaking long-range radar observation in 2003) had detected radio specular reflections indicating flat, possibly liquid, patches on the surface. But the lack of a glint in ISS imager implied that there was little if any exposed liquid.

Another puzzle raised by ISS was the structure of the haze. Close-up images of the north polar region showed many distinct layers of material, and it was not clear how these were organised. In addition, a 'detached' haze layer, distinctly above the main haze deck, was clearly seen at an altitude of 500km (it was best seen using the ultraviolet filter at 338nm), yet it had been observed by Voyager in 1980 at an altitude of 350km – an enormous difference. Were they even looking at the same thing? Or were there profound seasonal changes in the past (and perhaps yet to come in the future)?

On the outbound arc of Cassini's third orbit around Saturn, Cassini released the Huygens probe to make its historic descent through Titan's atmosphere. As Cassini continued outwards it had an opportunity (a bonus from the mission redesign to accommodate the receiver Doppler problem) to observe the satellite Iapetus.

The mass of Iapetus was not precisely known, and Huygens and Cassini passed close enough that their trajectories were nudged slightly by Iapetus's gravity. Since the mass was uncertain, so was the nudge, and this in turn expanded the uncertainty in the Huygens probe's arrival point (and most important, its hypersonic entry angle) at Titan. In fact, at a rather late stage this first Iapetus flyby had to be opened from a range of 64,000km to 126,000km to reduce this perturbation uncertainty to an acceptable level. Nonetheless, the flyby provided grandstand views and an opportunity to see this mysterious satellite's surface in unprecedented detail, the best resolution of 0.8km/pixel for ISS being far better than that of Voyager. These images revealed a curious ridge along Iapetus's equator, making the moon look like a walnut.

Then, as 2005 began, world attention shifted from Cassini to Huygens, and project scientists flew to the European Space Operations Centre in Darmstadt, Germany, for what would be for many the highlight of the entire mission.

LEFT A 2x2 ISS NAC mosaic of Iapetus, mostly of the dark Cassini Regio (named after the astronomer who first noted the asymmetry in the moon's brightness as it orbits Saturn) which has a reflectivity of only 5% or so. Arcing along the equator is a narrow ridge that stands 13km high. *(NASA/JPL/Space Science Institute)*

Chapter Seven

Probe mission

The Huygens descent would be a landmark in planetary exploration – unveiling close-up the last unknown world in the solar system with a major atmosphere. But this pivotal event relied on systems working correctly, first time: there would be no chances to fix any problems.

OPPOSITE Artist's impression of the Huygens probe on the surface of Titan, its deflated stabiliser parachute at upper left. A small area around the probe has been discoloured by the removal of dust by the aerodynamic wake, and the scene is littered with rounded cobbles. *(ESA)*

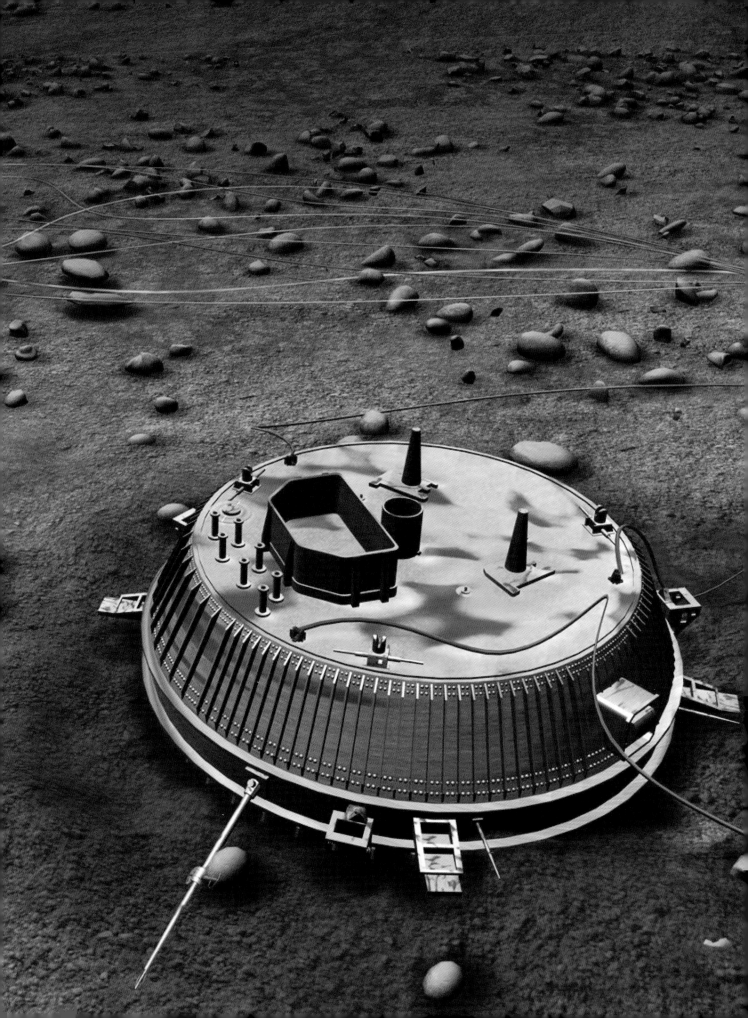

Descent into the unknown

Having released the Huygens probe on 24 December 2004 on a trajectory which would intercept Titan, Cassini performed an orbiter deflection manoeuvre (ODM) four days later in order to divert its own path away from that moon, and then a clean-up manoeuvre on 3 January 2005 set up the flyby for the probe relay operation.

Between separation and the ODM, Cassini attempted to snap the receding probe (in fact a similar parting shot by a dedicated camera on ESA's Mars Express mission the previous year was the last that was seen of the Beagle 2 probe until it was found on Mars 12 years later). Imaging of the probe would shrink the uncertainty envelope in which Huygens was delivered to Titan from an estimated 58x11km to 27x6km.

Some 12hr after release, at a distance of 18km, the probe was present at the edge of the centre frame in a 5x5 Wide-Angle Camera mosaic, confirming a safe delivery. Although by 52hr after separation it was so far away as to be smaller than 1 pixel in the WAC field, it was still able to be resolved in an NAC mosaic.

The Huygens probe was released on a ballistic trajectory that took 20.3 days to reach Titan. During this time, it was dormant apart from the three redundant timers which were counting down to a specific time programmed to end 4hr 23min prior to the predicted entry into Titan's atmosphere. (In the event that the timers failed, 'g-switches' would sense the deceleration and would activate the probe systems in order to recover at least part of the scientific mission during the descent.)

During this three-week coast phase, which in fact began with Huygens arcing out away from Saturn towards Iapetus before falling back towards Titan, alone in deep space without any warmth from its own systems or from Cassini, temperatures inside the probe were kept acceptably warm by the radioactive decay from

BELOW A remarkable NAC image of the Huygens probe receding into the darkness two days after separation from a range of 52km. The irregular pattern of brightness is due presumably to sunlight glinting off the multilayer insulation blankets on the back of the probe. (NASA/JPL)

38 plutonium Radioisotope Heater Units (RHU). The orientation of the probe was kept stable by a gentle (~7rpm) spin imparted by its spring-loaded release along spiral rails. This was to ensure that it penetrated Titan's atmosphere with the apex of its conical heatshield pointing forward.

At the appointed moment, battery power was turned on and the computers, their sensors (accelerometers, and later in the descent the radar altimeters), and the scientific instruments were energised according to a pre-programmed sequence. At 04:41:33 UTC on 14 January 2005 the probe 'woke up' as planned and its systems warmed up – the warm-up time had been extended from the original design to 4hr, to slightly improve the bit synchronisation associated with the relay design flaw.

The probe relay receivers onboard Cassini were powered on from 06:50:45 to 13:37:32 UTC. The probe arrived at the 1,270km interface altitude on the predicted trajectory at 09:05:53 UTC, just a few seconds ahead of the expected time. In fact, the sensitive accelerometer (a Sundstrand QA-2000) of the Huygens Atmospheric Structure Instrument detected the first tugs of atmospheric drag a few seconds earlier, at an altitude of about 1,300km.

As the probe penetrated the deeper, denser layers of Titan's atmosphere, the drag and heating built up. It would have been enveloped in a glowing hot wake, tortured into incandescence (up to 10,000K, which is hotter than the surface of the Sun) by the strong hypersonic shock wave created by the blunt shape of the heatshield. Several large telescopes on Earth were observing Titan in the hope of detecting the 'meteor trail' from the probe, but nothing was seen.

 The multi-layer insulation blankets of metallised Kapton that had kept the probe temperatures tolerable during seven years in space were ripped away by the flow. The drag built up to a maximum of 121.217m/s^2 (12.3g) shortly after the peak in heating rate.

The heatshield was not instrumented to measure loads, but post-flight estimates indicated the stagnation point (the apex of the heatshield) was subjected to about 45W/cm^2 of direct heating by the gas flow (i.e. convection) and about another 45W/cm^2 due to radiation from the glowing shock layer. Uncertainties

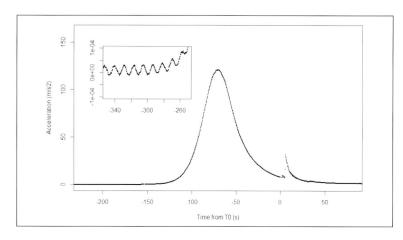

in the aerothermodynamic calculations, and indeed in the composition of Titan's atmosphere itself, when the probe was being designed 15 years earlier led (prudently) to much higher estimates (140W/cm^2), and so the heatshield was very conservatively designed; tests had shown that the AQ60 material should be able to handle 250W/cm^2.

The entry deceleration was recorded not only by the HASI accelerometer, but was also checked eight times per second by three redundant accelerometers (CASU) wired to the two probe computers. When two of them indicated a deceleration of less than 10m/s^2 after passing the peak value, the CDMU declared 'S0', triggering the parachute deployment sequence.

The descent phase of the mission was defined to begin at 'T-zero', set to be 6.375sec after S0 detection, with the closing of the relay that fired the pyro cartridge on the pilot parachute mortar. This event could be discerned as a small blip in the acceleration record, and in the next 2.5sec the pilot parachute inflated and stabilised.

Then the back cover pyros were fired to allow this to be pulled away and drag out the main parachute, whose lines became taut after about 1.4sec. And then 32.5sec after t0, once the probe had settled into a steady descent under the main parachute, the pyros were fired to permit the heatshield to fall away.

After its already epic journey, Huygens was ready to begin its measurements in earnest, and 44sec later power to the probe transmitters was switched on, launching news of the probe's arrival into space at the speed of light, 300,000km/s.

The probe's two radio transmissions were picked up about 0.25sec later by Cassini,

ABOVE The entry acceleration recorded by the HASI instrument, peaking at just over 12g (t0–70sec). A close inspection (inset) of the pre-entry data indicates a tiny periodic signal due to slight coning of the vehicle. The bump just after 0s is the deployment of the pilot parachute, followed by the sharp spike of the main chute inflation. A small bump 30sec later is the release of the heatshield. *(Author)*

71,500km away, and the receivers began to lock on. Unfortunately the Channel-A signal at 2,040MHz with left-hand circular polarisation would not lock because the selected precision reference oscillator had not been powered up, but Channel-B at 2,080MHz with right-hand circular polarisation started to be received correctly within 2sec, feeding 8kbps of data to Cassini's Solid State Recorders.

At this time, Cassini was 1,207 million km (8.07 AU) from Earth and the one-way light-time was 67min 6sec. The 100m Byrd Green Bank Telescope (operated by the National Radio Astronomical Observatory in West Virginia, USA) picked up the signal. There was far too little energy in one eight-thousandth of a second to permit discrimination of a 0 from a 1, so the datastream could not be extracted from the signal, but averaging the spectrum over several seconds confirmed the existence of the transmission (and would later allow measurement of Titan's winds via the Doppler shift on the precisely transmitted Channel-A signal).

The spectrum plot showing the clear spike of the signal from the probe was immediately faxed from the Green Bank to the European Space Operations Centre in Darmstadt, where anxious Huygens scientists greeted the news with jubilation.

It would be another couple of hours before the probe completed its mission, and 1hr of margin was built into the sequence to allow for timing uncertainties, and then it would take some time for Cassini to slew round to point to Earth and start to relay the data, a process which would itself take some hours because the probe data would be transmitted several times to ensure safe reception even if there was bad weather at the ground station.

Meanwhile the Huygens probe was descending and taking data. At t0+48sec, current pulses were sent to electromagnets to release the HASI booms, which swung out into the airflow. Over the next 18sec, caps sealing the inlets to the ACP and GCMS instruments were opened, allowing their sampling to begin, and a contamination cover protecting DISR was released, enabling this to take its first image at an altitude of 143km.

Later, as the probe dangled and swayed gently under the main parachute, 900sec after

t0, a final set of pyrotechnic charges were fired to sever the bolts holding the main parachute bridle. As it accelerated in free-fall in local gravity, a bridle leg pulled out a bag to deploy a small 'stabiliser' parachute, attached to its own bridle. This would permit the probe to reach the ground in just over 2hr, while Cassini was still available to relay its signal.

In the faster descent, the probe was buffeted rather more than expected, producing rough oscillations on its tilt sensors and causing a measureable blur on a handful of the DISR images. Although some people had expected a simple, slow pendulum motion for much of the descent, a faster 'scissors' mode in which the parachute and probe oscillated in opposite phase appears to have prevailed because the damping of the parachute motions was not as large as had been expected (although, in fact, the SM2 test had shown the same behaviour). Analysis showed the motions at about 20km altitude to have characteristics that could be attributed to ambient atmospheric turbulence. In contrast, below 10km, the descent was very gentle.

As the data was played back by Cassini, a chosen few scientists and engineers in the main control room could see hints of what occurred from some status flags extracted from the telemetry stream, prior to the full science dataset being assembled. There were mode changes in experiments indicating that the probe had reached the surface (or at least, believed it had reached the surface) and was continuing to function. The very length of the received telemetry record implied the probe had survived for an extended period, and indeed while the 64m Parkes radio telescope in Australia had been queued up to listen for the signal from the probe, when it was realised that Huygens was still operating some 4hr after arrival, additional monitoring had to be hastily arranged further west as Earth's rotation took Parkes out of view of Titan.

Meanwhile, on the voicenet at ESOC came an ominous query: "At what time was the RUSO power-on command?"

The DWE experiment to monitor Titan's winds by measuring the Doppler shift on Channel-A relied on precise knowledge of the probe transmission frequency, fixed by the TUSO onboard, and comparing its received signal on Cassini with a receiver oscillator (RUSO) onboard. But it was immediately realised that while the receiver had correctly selected the RUSO over a standard, less precise unit, somehow the command sequence running on Cassini had failed to include the instruction to power on the RUSO. This disastrous error was in principle an easily avoidable one, but somehow the omission slipped through the checks.

Not only did this mean there was no Doppler measurement onboard, it meant that data from Channel-A, transmitted perfectly from the probe, wasn't received and recorded – it was lost forever. Eventually the goals of the DWE experiment would be largely recovered by analysis of the Channel-A signal recorded by radio telescopes on Earth, but the loss in accuracy and detail was significant.

Although all the 'critical' science data was deliberately transmitted on both channels for redundancy, for other data the scientists had exploited having two channels to send nearly twice the total amount of data originally planned, hence the loss of Channel-A meant the final dataset had some keenly felt omissions – not least half the number of images that would otherwise have been sent.

The probe itself was ignorant of these problems, and operated more or less like clockwork. The various instruments made their measurements as the probe descended. But there were a few anomalies. One ion source in the GCMS instrument failed – several electrical transients were noted around that same time, although whether these were a common cause (such as an electrical discharge or a turbulent 'bump') or were caused by the ion source isn't clear. Some peculiarities of the HASI electrical data, and puzzling aspects of the motion under the parachute may even hint that for at least part of the descent, one of the booms may have failed to latch into place.

Although the ACP instrument acquired and pyrolysed its samples, the chemical analysis of the pyrolysis products would later show less than had been hoped for. It appears that during the first analysis it made, the silicone elastomer seal on the oven was too cold to seat properly (like the O-ring seal between solid rocket booster segments that failed in 1986, causing

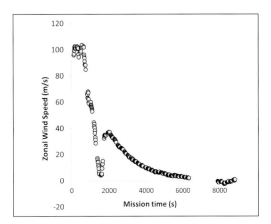

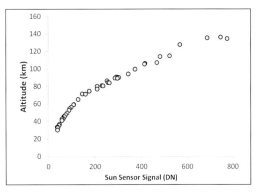

LEFT The zonal wind speed recovered from the ground-based Doppler record has several gaps that would not have been present had the Doppler measurement by Cassini been successful. The strong super-rotating winds (like a jetstream) at high altitude fell off abruptly in a shear layer at about 80km, before increasing again. A small westwards flow was encountered in the last few kilometres of the descent. *(Author)*

ABOVE Light levels faded as the probe descended through the haze. DISR measurements at different angles and wavelengths provided insights into Titan's 'anti-greenhouse' effect due to haze absorption and scattering. The signal from the Sun sensor fell rather more steeply than had been expected, possibly because the sensor got too cold and ceased providing useful data below 30km, although this ultimately had minimal science impact. *(Author)*

BELOW Temperatures inside the probe (top curve, DISR electronics) were roughly at room temperature throughout descent – electrical power dissipated inside caused temperatures to rise slightly for the first hour, before the thicker, colder atmosphere (HASI air temperature, bottom curve) began to take hold. After 8,870sec, now on the surface, the air was no longer rushing past and therefore the rate of cooling declined. Most instrument temperatures (e.g. DISR optics) were somewhere in between these two limiting curves. *(Author)*

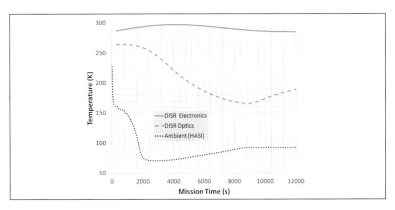

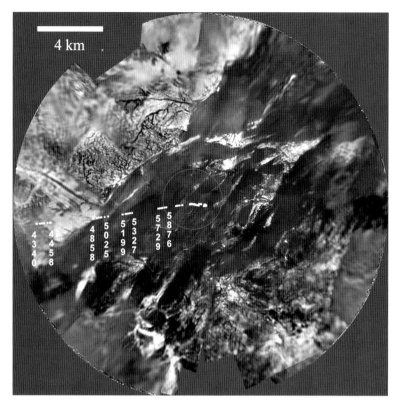

4 km

ABOVE A mosaic of DISR images centred upon the landing site (some parts are seen more clearly than others) with the ground-track (which makes a slight loop at the end) shown in white with mission-time markers. The probe landed in the complex dark area, believed to be a former riverbed. To the north-west are branching dark stream channels in a bright highland area. *(U. of Arizona/ESA/NASA)*

BELOW The Huygens mosaic reprojected to show the bright highland region (only about 100m above the dark plain). Two near-horizontal stripes in the upper right (~20km north of the landing site near centre) are sand dunes and were the key features in matching the DISR mosaic with radar imaging from Cassini. *(E. Karkoschka/U.Arizona)*

the loss of the Space Shuttle Challenger). However, the analysis did confirm that the haze particles contained nitrogen – as some models of the atmospheric chemistry had suggested.

The radar altimeters were switched on 32min after t0, corresponded to an altitude of about 60km. They fed altitude measurements to the onboard computers, which filtered and compared the measurements to the predicted altitude, to exclude erratic measurements at high altitude and to provide reliable measured altitude information to the payload instruments.

This allowed for optimisation of the measurements obtained during the final portion of the descent. The DISR sequence was adjusted to measured altitudes below 10km and its lamp was switched on at a height of 700m to provide 'natural' light. The HASI and SSP instruments were set to their proximity and surface modes at low altitude above the surface.

The probe reached the surface at the expected vertical speed of about 5m/s precisely 8,869.76sec (about 2.5hr) after t0, at the 'long' end of the expected descent duration, indicating perhaps slightly higher drag on the parachute than had been expected.

The moment of impact was captured by the SSP penetrometer, as well as by the accelerometers on the HASI and SSP

BELOW Crème brûlée? The 1/20th of a second signal from the penetrometer indicated a deformable surface without large gravel grains (at that spot, at least). The spike at the beginning of the record was initially interpreted as a possible crust, but it seems more likely to have been a pebble. A small ramp-up at sample 50, before the spike, may be a layer of fine dust on top of the pebble. *(Author)*

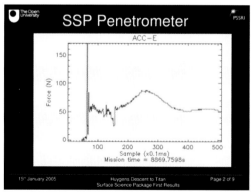

experiments – the Doppler shift on the radio signal also abruptly changed. The impact was a thump rather than a crash (5m/s is only 10mph), and the peak deceleration recorded by the fast-sampled (500Hz) and rigidly mounted SSP impact accelerometer was 18g, consistent with a somewhat soft surface into which the probe penetrated about 12cm.

The HASI accelerometer at the centre of the experiment platform recorded only a 12g peak deceleration, muted somewhat by the flexing of the probe structure and its lower sample rate (200Hz). This sensor did, however, record the subsequent motion, indicating that the probe had a ~0.4sec bounce upwards.

Detailed forensic analysis of data from DISR and other sensors showed that the probe probably skidded sideways out of its dent – much as the SM2 test unit had done in the Swedish snow – and wobbled for a couple of seconds before finally coming to rest.

Diagnosing what happened to a vehicle in an unknown alien environment a billion miles away based on only a few dozen measurements is an exercise with some ambiguities, but an interpretation of DISR data suggested the stabiliser parachute drifted to the south-south-east of the probe in light (0.3m/s) winds, and that a cloud of dust was briefly kicked up upon impact by the aerodynamic wake of the probe.

The 512 bytes of data from the penetrometer, recorded in the 1/20th of a second around the moment of impact received prompt examination, and its data were readily interpretable as indicating a soft surface overlain by something that caused a higher force reading; a more detailed examination later also teased out a hint of a soft dust layer before (above) whatever caused the spike.

The off-the-cuff 'instant science' interpretation at ESOC was that Huygens had landed on a surface with the texture of 'packed snow or damp sand' overlain by a harder material. This prompted the SSP impact analysts to draw a comparison with crème brûlée – an analogy which immediately caught the press's attention. When the DISR images became available, however, it seemed more likely that the spike had been caused by an impact with one of the pebbles of ice that littered the landing site.

It was the post-landing DISR imagery of a level plain littered with rounded cobbles and pebbles which probably defined the Huygens results more than anything else. And yet we had no right to expect such a stunning view.

Given the vagaries of the mission, the probe could have been smashed on a hard surface; or tumbled into a gully, blocking its radio signal; or the parachute could have draped itself over the camera head. But we were lucky, and none of these things happened; DISR just kept on taking and transmitting data until the batteries expired. In fact, there were ~100 near-identical image sets, since the field of view was fixed at about 13° west of south taken from a point 48cm (about knee-height) above the surface, and little in the scene changed.

In the hours and days following the probe descent, an eager media and public focused their attention on the images taken on the surface. The initial plan was to release three images to the media on the night of the encounter. A young computer staffer set up a web page at the University of Arizona to make the chosen images available. The server promptly received a torrent of traffic. The three links on the web page with the preliminary captions pulled up the three chosen images.

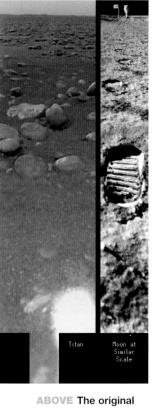

ABOVE The original DISR image triplets (left) had to be re-scaled, and 'flat-fielded' – corrected for photometric effects like vignetting (dark edges) and the 'chicken-wire' pattern caused by the fibre optic bundle (middle) – to accurately represent the unexpected scene at Titan's surface and transform our perspective on this world. The panchromatic (grey) view was tinted authentically using data from the upward-looking visible spectrometer of DISR and montaged with Apollo data at the same scale at right. *(Author, from elements by C.See and E. Karkoschka/U. of Arizona)*

So far so good. But the directory in which these images were stored actually contained *all* the images, and it didn't take the enterprising denizens of the internet long to realise that by simply changing the image number in the web address they could access the other images too. Word of this miracle spread fast, and soon amateurs everywhere were speculating about what the tiny images showed, and were trying to mosaic together the image jigsaw.

Although the DISR team was initially horrified by this error, the eagerness with which members of the public discussed the images and skilfully processed them demonstrated the high interest in the mission, and of course the team was not really 'scooped' of the 'hard' science that depended upon calibration and detailed analysis.

Surface changes

In broad terms, very little happened during the 75min that Huygens spent on the surface of Titan. The probe cooled down by several degrees, the ~300W of power dissipated by its systems internally being slightly overpowered by heat leaking through its insulation to the ambient environment.

Some of the heat from the probe might have sweated material out of the damp surface – an ultrasound sensor on the SSP ceased sensing

blips after 20min or so, perhaps because of muffling vapours, and the electrodes of the HASI mutual impedance experiment saw an increase in the electrical resistance of the surface material after 11min.

A couple of spots received particularly strong heating. First, the inlet of the GCMS was warmed by a 5W heater to prevent condensation in the inlet pipe during the descent through the atmosphere. After landing, this instrument saw elevated amounts of methane, ethane, acetylene, and carbon dioxide as the inlet tube warmed to an estimated 110–130K. In addition, a small patch of ground ~20cm wide was heated by the 20W surface science lamp of DISR, its heat flux of ~200W/m^2 being hundreds of times stronger heating than noontime sunlight on Titan. This heat may have been responsible for driving off methane vapour, some of which appears to have condensed on the probe: in one of the ~250 images taken on the surface there was one frame that showed a circular blob that seems to be best explained by a 4mm-wide drop of methane that presumably formed as a dewdrop on the camera baffle. One study claimed to detect some drifting fog on the horizon, but the relevant image brightnesses are right at the noise level of the camera and hence are suspect. Inspection of the tilt sensor and accelerometer data, as well as the camera images, shows the probe settled a tiny amount (about 0.3°) during the time it operated on the surface.

While Huygens was on the surface, Cassini was getting progressively closer to Titan, but the strength of the radio signal it received from Huygens actually got lower because Cassini's elevation as seen from the probe was getting progressively lower, and therefore the line of sight was in the weaker part of the wide beam from Huygens's antennas.

Whereas the signal strength had varied rapidly and periodically during the descent, as the spinning of the probe swung the lobes of its beam pattern around like a disco ball, the strength was much more constant on the surface. But the record did show three large scallops that were initially puzzling. These proved to be due to 'multipath' – the interference of the shallow upward direct signal from Huygens to Cassini with a second signal, leaking shallowly downwards from the

BELOW The argon-40 signal in the mass spectra from the GCMS gave a major clue to the origin of Titan's atmosphere as having being captured in the form of ices. Overall the atmosphere composition was simpler than expected, with few large compounds detected (and by implication, perhaps the gas chromatograph wasn't a necessary payload). The composition of the surface was rather richer, with traces of cyanogen (C_2N_2) and Benzene (C_6H_6) and perhaps other compounds. *(NASA)*

probe and bouncing off the surface. As Cassini passed through this interference pattern, the intensity went up and down, and in fact the curve was very sensitive to the geometry, giving a precise measurement of the height of the antenna above the ground, and indicating that the probe was sitting flat on the ground and thus had bounced or skidded out of the dent it made at impact (the same sensitivity is why small changes in the position of a cellphone can change the quality of the signal).

Some 50.7min after impact, the signal level dropped below the 3.3dB threshold of the radio receiver and some data packets were lost as the rate of bit errors exceeded the capacity of the Reed-Solomon decoding to correct. However, the signal came back up again and a further 16min of data were received.

The last packets of Huygens data were recovered by Cassini after 218min of descent. Cassini's receiver still detected the carrier for another couple of minutes, until the link was lost at 220min (12:50:20 SCET, or 72min after landing), about 1min after Cassini dropped below the geometric horizon. By that time the spacecraft was at an elevation of minus 0.43° relative to the probe, implying the radio signal was slightly refracted by the atmosphere. Ground-based telescopes continued to detect the carrier for another 52min or so, and it is calculated from battery telemetry and the ~250W power consumption that the probe probably continued to transmit for another hour, until it finally fell silent between 14:52 and 15:16 UT, some 3.5hr after landing. This performance far surpassed expectations.

Not everything had worked perfectly. In addition to the RUSO commanding error that lost all the data on Channel-A, the two radar altimeters both made 'false locks', reporting about half of the true altitude from about 25km and 17km. Even when Radar B recovered to the true altitude at about 22km it was 'outvoted' by the probe's onboard logic, which had seen consistent (but incorrect) reports from both altimeters. But fortunately Radar A recovered just in time, before DISR and SSP triggered their near-surface operation modes. A total of 608 images were received by Cassini, of which 360 were taken during the descent and the remainder from the surface – out of a total of

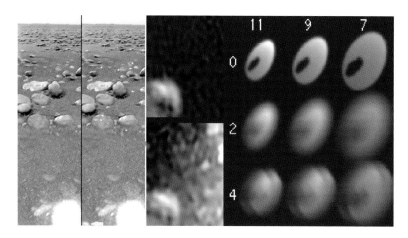

ABOVE A painstaking analysis and comparison with an 'averaged' image (left) showed that most of the small transient features in the post-landing DISR images were cosmic ray spikes on the detector, or artefacts of the lossy data compression – except one on image 897 in the bottom (middle-left panel, yellow arrow) just to the right of the bright spot caused by the surface science lamp. The arcuate shape seen by blowing up the image and subtracting the background (middle-right) could be replicated using a model of light from the lamp refracted by a methane droplet a couple of millimetres in diameter and some 7–11cm in front of the camera. *(E. Karkoschka/U. of Arizona)*

HUYGENS MISSION TIMELINE ON 14 JANUARY 2005

Activity	Time (UTC)	Mission time (t-t0)
Probe power-on	04:41:18	-4:29:03
Probe support avionics power-on	06:50:45	-2:19:56
Arrival at interface altitude (1,270km)	09:05:53	-0:04:28
t0 (start of the descent sequence)	09:10:21	0:00:00
Main parachute deployment	09:10:23	0:00:02
Heatshield separation	09:10:53	0:00:32
Transmitter ON	09:11:06	0:00:45
GCMS inlet cap jettison	09:11:11	0:00:50
GCMS outlet cap jettison	09:11:19	0:00:58
HASI boom deployment (latest)	09:11:23	0:01:02
DISR cover jettison	09:11:27	0:01:06
ACP inlet cap jettison	09:12:51	0:01:30
Stabiliser parachute deployment	09:25:21	0:15:00
Radar altimeter power-on	09:42:17	0:31:56
DISR surface lamp on	11:36:06	2:25:45
Surface impact	11:38:11	2:27:50
End of Cassini-probe link	12:50:24	3:40:03
Probe support avionics power-off	13:37:32	4:27:11
Last Channel-A carrier signal reception	~14:53	5:42:39
by Earth-based radio telescopes	16:00 (ERT)	-

1,211 transmitted, of which 722 were taken during the descent.

A dedicated Descent Trajectory Working Group (DTWG) of Huygens scientists worked to simulate the Huygens trajectory on the basis of the delivery ephemeris and various probe datasets. They determined a landing latitude of 10.3° (±0.4°) south and a longitude of 167.7° (±0.5°) east. When the probe landed, the Sun was ~32° above the horizon, at an azimuth of 115.5° (at the beginning of descent, the solar elevation was around 39°, the angle dropping in part due to Titan's rotation and in part due to the westward drift of the probe in the wind).

The second column of the table on P157 gives the time in UTC (for the probe), while the third column gives the time relative to t0, where t0 is the official start of the descent associated with the pilot parachute deployment event.

Putting a spin on the probe

During its 2.5hr parachute descent, the spin of the Huygens probe was designed to pan around the field of view of its side-looking camera (and more importantly, but less widely appreciated, to optimise the measurement of light scattering by the haze). A separate spin requirement was that the probe be spun for gyroscopic stability in order to ensure that it would make its hypersonic entry 'head-first' with a minimal angle of attack.

The spin rate during the coasting phase was determined to be 7rpm – slow but enough given the probe's large moment of inertia to prevent small nudges by solar radiation pressure, or magnetic or other effects from nudging the spin axis too far in the three weeks of passive coast. The spin was achieved on separation from Cassini by pushing

Huygens away on three spiral rails using rollers and springs. As the probe was pushed away, special low friction connectors of the umbilical were yanked out. Further evidence of the separation event was inferred from the measurement by the inertial reference units as Cassini was torqued in the opposite direction.

After a short period to allow Huygens to reach a safe distance, Cassini's AACS took over to cancel this induced tumbling. In fact, the thrusters were disabled for 5min in order to let the fuel slosh die down in advance of beginning this recovery action. The thrusters took 168sec to bring the rotation down to zero, and then made an 11.6min slew to point the HGA at Earth to report the separation.

The best estimates of the mass properties of the two vehicles indicate that the separation velocity was 0.34m/s and the probe was spinning at 6.7rpm (well within the specification of 0.31–0.43m/s, and 5–10rpm) with Huygens pointed within 0.2° of the desired vector for entry.

The departing probe was also seen detected as a decaying 7.4 cycle/min oscillating signal on Cassini's magnetometer owing to the magnetic components on the probe – the higher frequency was because of Cassini's opposite rotation.

When the probe was switched on after quietly coasting and the HASI accelerometer began recording in preparation for entry, its data showed a 0.085Hz oscillation with an amplitude of 1.8µg. This indicated a tiny coning motion caused by the angular momentum imparted upon separation not being quite aligned with the axis of maximum moment of inertia of the probe. This coning had a rotation of 7.04 ±0.16rpm.

Small torques generated from uneven aerodynamic effects on the probe during atmospheric entry (e.g. when the multi-layer insulation blankets tore off, or the slightly offset

RIGHT In the intended plan, the downward-looking high-resolution camera on the DISR instrument would produce several circular mosaics, forming a 'teardrop' coverage as the probe drifted east in the zonal winds as it descended. In the event, the distribution of azimuths of the images was much more random, and the probe drifted rather less than expected. Furthermore, the surface only became visible at low altitude, due to the ubiquity of the haze. (U. of Arizona)

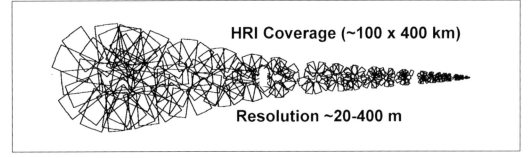

HRI Coverage (~100 x 400 km)

Resolution ~20-400 m

pull from the pilot parachute as it inflated) may have altered the spin rate very slightly; the first spin rate data measured by the RASU unit on the probe indicated 6.99rpm.

The spin rate desired for the descent was 1–5rpm, to enable the DISR camera to pan around. In order to achieve this the probe's rotation was decoupled from the parachute by a swivel in the riser, and a set of small vanes around the perimeter of the probe developed a torque to induce rotation. Although previous probes (Russian Venus probes, and NASA Pioneer Venus and Galileo) had used spin vanes before and seem to have functioned adequately, their performance has not been critically examined. Only with an imager have the attitude dynamics of a descent probe required close attention.

The spin vane configuration for Huygens had many vanes with a very shallow setting angle of just 1–3°. In essence, the vane angle defined a helical path which the probe perimeter would try to follow – a helix with a pitch of 2° or (1/30) meant that a vane 0.65m from the axis of the probe would make one full turn (2 pi*0.65 ~ 4m) every 120m of descent. This 'ideal' spin rate profile in rpm was just a scaled version of the descent speed profile. In fact, in executing a spiral at exactly the setting angle, a flat aerofoil like the vanes would be at zero angle of attack and so develop zero torque. Hence for the vane torque to balance the small friction in the swivel, the equilibrium spin rate was fractionally less than that corresponding to the setting angle. Moreover, the equilibrium spin profile would only be realised if the vanes had infinite area and torque. In practice, the small vanes would require a finite time to accelerate the heavy probe, so the real spin rate profile was a sort of smoothed version of the descent speed.

Hence the expected spin profile would begin at whatever rate resulted from the release and entry, in the range 5–10rpm, and tend towards the scaled (slow) descent speed under the main parachute. However, when the main parachute was jettisoned and the rate of descent increased, the probe would be spun up again to a maximum value from which it would slowly decline as the rate of descent diminished in the thicker lower atmosphere.

The probe measured the rotation rate in real time by interpreting date from two radially mounted accelerometer sensor units (RASU) as centripetal acceleration due to rotation. Although this was a somewhat imprecise approach (consideration was given to using HRGs, which would have been much better, but these were unfamiliar to the European team at the time the probe was designed and the RASU used accelerometers that were already qualified and used on CASU) the requirements were not exacting and this design would be adequate.

So much for the theory. During the probe descent the spin rate profile indicated in the RASU data (which was just the spin rate without any indication of direction) began at about 7rpm as expected, then declined as expected, or perhaps slightly faster. Then the spin rate increased again, as expected, and then declined with a few wiggles. A casual inspection of the curve suggested it was a bit rough, and it was odd that the spin rate became virtually zero before climbing back up, but broadly the shape was as expected.

But something didn't add up. The solar data and imagery from DISR were not following the expected pattern, and images that should have overlapped didn't. There were also some puzzling features in the SSP tilt sensor data (one sensor, aligned radially, responded to the centripetal acceleration like RASU). One DISR scientist raised the question of whether the probe could have spun the wrong way. This was regarded as an outrageous suggestion!

The key dataset to unravel the mystery was not one for which there were plans for detailed analysis. The Automatic Gain Control (AGC) in the Huygens receiver on Cassini was reported in the PSA telemetry eight times per second, providing an indication of the received signal strength. Obviously it was available only on Channel-B because the Channel-A receiver had not locked on. Since the pattern of transmitted power from the probe antennas was not a perfectly smooth function of elevation and azimuth, the rotating probe beamed different levels of power in different directions, like a disco ball. The received signal strength at Cassini was like a circular cut through this emission pattern, and therefore defined a 'heartbeat', repeating once per rotation. And because the pattern was not perfectly symmetric, the trace of signal strength could be modelled using ground

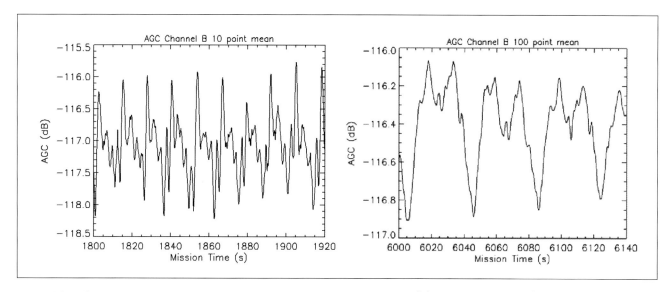

Snapshots of the automatic gain control reading on the Cassini receiver, indicating signal strength from Huygens. A regular but asymmetric 'heartbeat' pattern showed that the spin rate decreased and then reversed its direction! *(Author)*

BELOW The expected (dashed) and reconstructed (solid) spin rate profile, showing that the spin reversed, and changed rather more rapidly than a steady-state condition might suggest. *(Author)*

tests of the antenna pattern to infer the rotation direction as well as the rotation rate.

This analysis showed that although the probe was initially spinning the right way, it had quickly slowed down to zero and then continued that trajectory to spin up the wrong way!

This discovery prompted a review of data from the SM2 balloon drop test a decade earlier, for which the possibility of a reverse spin had not been considered, so the sign

of the gyro sensors on that test had not been examined. And because the probe was released from rest, the zero-crossing behaviour in the actual descent did not occur to draw attention to the spin rate history. However, study of the Sun position in the upward-looking video camera showed that indeed the SM2 test had spun the wrong way too!

The spin rate profile during SM2 wasn't quite as expected, but it was reasonably close. The geometric accommodation of the spin vanes on the probe Descent Module (in the limited space between the probe hull and the heatshield) meant that the number and area of the vanes was limited. However, the setting angle could be changed (in fact by bending the vanes slightly with a pair of pliers) and it was changed after SM2. Perhaps, had the SM2 spin anomaly been recognised, such tweaks would have been enough, but perhaps not much could have been done except to modify instrument software to take uncertain or reverse rotation into account.

In the end, the degradation of the DISR dataset due to the spin anomaly was modest, because sufficient angular sampling was obtained to construct image mosaics and light scattering functions as planned, but achieving these results took considerably more work than expected. Of course, had the Channel-A data not been lost as well as the spin being the wrong way, the number of images and data points would have been larger – on the one hand increasing the workload further but on the other hand filling in many more gaps.

Whatever caused the reverse spin had to be

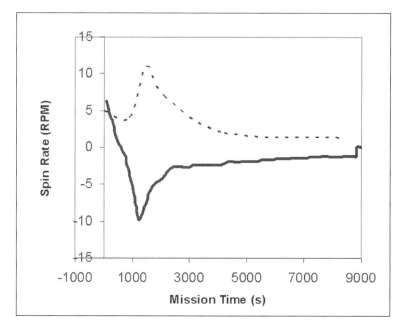

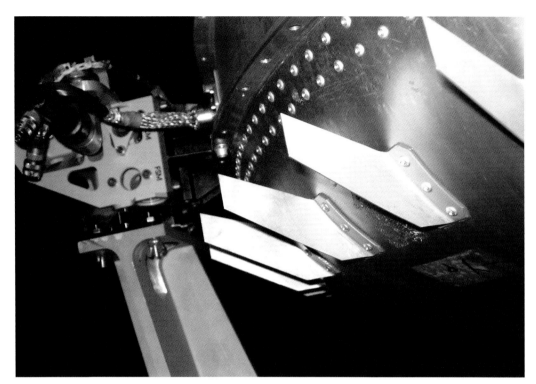

common to the SM2 test and Huygens in-flight. An obvious culprit was the SEPS separation mechanisms, the three bulky brackets that affixed the Descent Module to the heatshield. The torque these might develop had not been evaluated, and it was only during wind tunnel tests (with a probe model mounted on a special air-bearing swivel) at the Von Kármán Institute in Belgium in 2014 that they were measured and found to provide an adequately strong torque to overcome the vanes.

But even this was not the full story. The SEPS, acting as reverse vanes, would still give a rather smooth profile. Yet the actual spin history was slightly irregular, implying the presence of a time-varying torque. Attention focused on the HASI booms, which were suspected of not properly deploying early in descent. Some of the mutual impedance data simply made no physical sense, if the deployed electrode geometry was assumed. Also the SSP tilt data suggested the aerodynamic force on the probe was off-centre for a while. All of these indications were consistent with one of the HASI booms sticking, or at least not fully deploying, for several minutes early in descent, before finally latching into place (there was no switch to confirm successful latching into place).

All these intriguing little niggles apart, the Huygens mission to Titan was spectacularly successful and achieved its science objectives.

BELOW The interaction of the DISR software with the unexpected spin direction (and loss of function of the DISR Sun sensor at this point) meant the image acquisition was probably not optimal in some sense, but it was adequate to cover the landing area, and the available overlaps allowed stereo photogrammetry to estimate topography. (U. of Arizona)

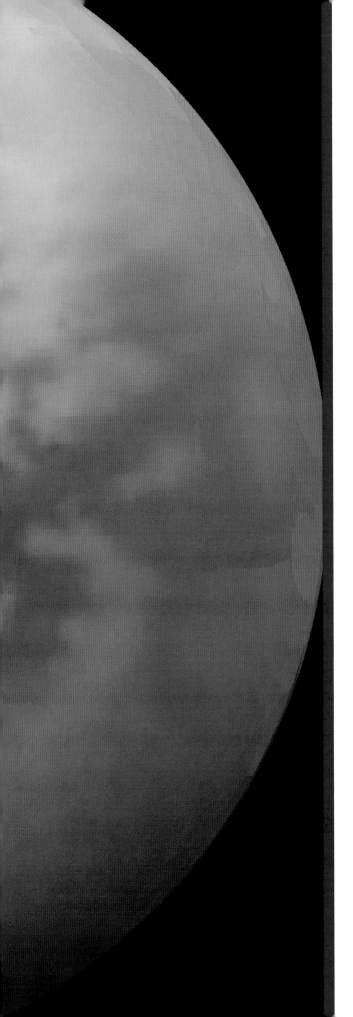

Chapter Eight

Ongoing discoveries

Day in, day out, for over 12 years, Cassini has been hard at work orbiting Saturn and observing the rich and changing array of targets in the Saturn system. The pace of its discoveries has been phenomenal, barely slowed by occasional glitches in otherwise well-oiled machines in space and on the ground.

OPPOSITE A false-colour infrared map of Titan from VIMS. The red patch at the lower left, Tui Regio, may be a lakebed. *(NASA/ JPL/U. of Idaho)*

ABOVE A close-up image of Iapetus's surface spanning about 55km, showing the irregular rounded patches of bright and dark material.
(NASA/JPL/SSI)

Major revelations

There is no space in this book to properly cover the spectacular discoveries made by Cassini after the Huygens mission. In over a decade, this multi-billion-dollar spacecraft and its arsenal of instruments have been trained on Saturn, its rings and magnetosphere, and Titan and the smaller icy satellites – a scientifically rich arena. Hundreds of thousands of images and spectra and vast amounts of other data

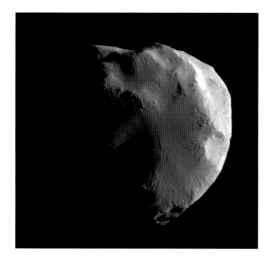

RIGHT Bizarre and unexpected flow patterns on the tiny moon Helene.
(NASA/JPL/SSI)

have been taken. Literally thousands of scientific papers have been written from this wildly productive mission. Here we can just mention a few examples.

In some cases the scientific discoveries occurred almost overnight, with key long-planned observations providing the missing piece of the puzzle. The close flyby of Iapetus in September 2007, for example, showed the relationship of the dark and bright material, allowing their distribution to be understood as a process of thermal segregation (areas with some dark material would become warmer, and so bright ice would tend to migrate to cooler areas, causing the dark spots to get darker and bright spots to brighten – a process that may also be at work on Hyperion, where dark material is found in what seem to be unusually deep crater pits). At last the mystery which Jean-Dominique Cassini had discovered, where Iapetus was brighter on one side of its orbit than on the other, was solved.

Other first-and-only looks have revealed the tremendous diversity of surface processes, such as strange flowery flow lobes on the tiny moon Helene, which orbits close to the rings.

In other cases, the complex truth has only emerged piecemeal, flyby by flyby. Radar image mapping of Titan, for example, can cover only about 1% of the surface at a time, and of the 126 flybys only about a third have made high-resolution radar observations. Some gaps were filled by low-resolution scans, but overall only about a half of Titan has been mapped at kilometre-scale-resolution or better.

Nevertheless, the distribution of the coverage in strips that spanned all latitudes and longitudes allowed some 'big picture' trends to emerge. Titan possesses seas of liquid methane and ethane, but only around its north pole, whereas only a single large lake exists at the south pole. VIMS data showed that this lake, Ontario Lacus, contains at least some ethane, and rings of bright material at its margins suggest that the lake was once wider but has dried up. The same brightness in other patches suggest the existence of dry lakebeds elsewhere on Titan.

Sometimes entirely unanticipated capabilities emerged as scientists gained a fuller understanding of the potential of the data from

Cassini. The way radar images are made from a mosaic of five narrow strips caused unseemly stripes when not corrected for the distance to the surface, but this 'blemish' permitted the engineers to estimate the surface height profile in an image. This development greatly expanded the coverage of topographic data over that which would have been possible using the instrument's altimeter mode alone.

In fact, topography on Titan has been studied at the millimetre level. As Saturn and Titan advanced into northern spring in 2009 and the Sun rose over Titan's north polar regions, the sunlight could be detected glinting off the mirror-like surface of the lakes. It is suspected that the lake surfaces have been dead flat because winds have been too gentle in the current season, but as winds pick up in the northern summer, the sea surface may become more ruffled (as hinted by small patches of sea surface that appeared to scatter radiation seen by RADAR and VIMS). A new academic field has emerged – extraterrestrial oceanography. It turns out that Titan's seas are rather radar-transparent, and Cassini's radar altimeter was able to detect an echo from the seabed, allowing scientists to model tidal- and wind-driven circulations in the seas.

Some of the most powerful insights arose when data from different instruments on Cassini were brought to bear. For example, a much better sense of the processes at work on Titan emerged when topographic and textural information from the radar were combined with indications of surface composition obtained by VIMS and ISS. Indeed, some of these processes

have been seen in action as the seasons progressed – during the equinox season in 2009 a massive rainstorm was observed in the equatorial zone by ISS. The ground was seen to have darkened, and then in the following months it progressively lightened up again, suggesting the ground had become wet and then dried out. Titan's hydrological cycle, which has carved the gullies and tumbled the cobbles seen by Huygens, was active as we watched.

Not only did Titan's weather change. So too

ABOVE A radar map of Titan's north polar region made in 2014, showing sprawling seas and dozens of small lakes. The mosaic includes several low-resolution areas and total coverage gaps. *(NASA/JPL/USGS)*

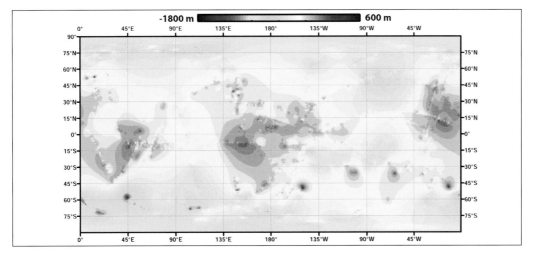

LEFT A Titan topographic map from radar data (areas not covered by radar are smoothly interpolated). *(Author)*

RIGHT Sun glinting off the surface of Jingpo Lacus on Titan, a methane-ethane lake. The mirror-like reflection means the lake surface was perfectly flat – winds had not yet picked up to form waves on the liquid surface. *(NASA/JPL/U. of Arizona)*

BELOW Extraterrestrial baythmetry. At the bottom is a radar mosaic of the second-largest of Titan's seas, Ligeia Mare at ~78° north, from flybys T25 and T28. The altimetry track from T91 is added in red. Near the middle, the echo shown at the top was obtained, the second bump showing the faint bottom echo appearing about 2μsec later than the main echo from the surface of the sea, suggesting a depth of ~150m. The bottom echo can be traced all the way across the sea (colour inset). *(M. Mastrogiuseppe/Cassini Radar Team)*

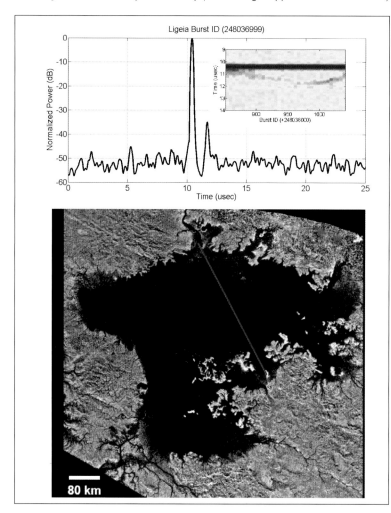

did that of Saturn. Subtle hues of the planet's haze changed as the sunlight distribution evolved, especially due to the movement of the ring shadow. In 2009, a huge storm system erupted in the northern hemisphere and provided a spectacle for several years. Heralded by an increase in radio noise due to lightning discharges, Cassini (and terrestrial telescopes) observed as the cloud system stretched into a band at that latitude. In addition to striking imagery, the spectrometers and even the radar radiometer on the spacecraft could monitor the evolving temperatures and composition as the storm dredged up air from deeper in Saturn's atmosphere.

The power of Cassini's broad payload was perhaps most notably brought to bear on tiny Enceladus, a small moon that was of interest for its high albedo (it reflects almost all the light that falls upon it) and because large portions of its surface are smooth. The possibility of geological activity was speculated upon during the formulation of the Cassini mission. In addition, this moon orbits close to the brightest part of the diffuse E-ring. Were the two related?

The first indications of some kind of emission or atmosphere were from Cassini's magnetometer, which noted that something bigger than Enceladus itself was blocking the plasma flow around Saturn. A UVIS occultation showed the presence of water vapour, not in a uniform atmosphere around the moon but over only a part of it. Then high-phase (i.e. backlit) views of the south pole by ISS showed dramatic plumes of (ice) dust blowing into space and feeding into the E-ring. VIMS observations helped to estimate the size of the particles in the plumes. It was also found that the extent of plume activity varied with the position of Enceladus in its orbit, suggesting that gravitational tides play a role in driving or at least in controlling them.

Radio science gravity measurements indicated that there is liquid in Enceladus's interior. A water ocean offered the prospects of prebiotic chemistry, and indeed measurements of the gas in the plume by INMS show water and organic compounds.

CIRS infrared measurements showed the glow of heat from the cracks out of which the plumes emerge. And the CDA detected salts and tiny grains of silica in the dust particles in the plume,

Map of Saturn's Moon Titan · June 2015

4000 km

ABOVE A map of Titan at 940nm wavelength by the Cassini ISS camera towards the end of the mission. The equatorial dark regions are tracts of dark sand dunes (as revealed by radar) whereas the polar dark spots are lakes. The kidney-shaped Ontario Lacus is at the bottom, near the south pole. *(NASA/JPL/SSI/USGS)*

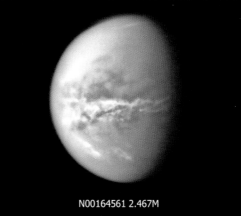

N00164561 2.467M

LEFT The 'telephoto lens' of Cassini's NAC could see features on Titan from a great distance (in this case, 2.5 million km in 2009) to study the weather, here showing that a bright cloud belt has appeared over the equatorial region. *(NASA/JPL/SSI)*

LEFT Shortly after equinox, a massive storm erupted in Saturn's northern mid-latitudes. Dredging up bright ammonia crystals, the storm was first detected by RPWS via radio emissions from lightning discharges. Over the following couple of years, the storm's wake grew to stretch around much of Saturn. *(NASA/JPL/SSI)*

showing that the water in Enceladus's interior is salty and its interactions with the rock take place at somewhat elevated temperatures.

As the significance of Enceladus was realised, every effort was made to develop better opportunities to observe it.

It was found that this remarkable little world has a significant influence on the overall Saturnian system. For example, traces of the water vapour from Enceladus find their way into Titan's upper atmosphere – where oxygen-bearing molecules would otherwise not be present. And charged particles originating from Enceladus spiral along magnetic field lines in the planetary magnetosphere to find their way onto Saturn itself, creating a 'footprint' in the auroral glow. Only with Cassini's formidable set of instruments could all these processes be investigated in an integrated way.

But beyond 'mere' scientific understanding of processes and origins, the goal of planetary exploration is to learn our place in the universe.

A dramatic view of Enceladus and its plumes as the south pole (at the top in this view) edges into winter darkness. The grooved surface is cut by cracks (sulci) from which the plume jets emerge. The jets are only easily visible at high phase illumination, a geometry that the Voyagers did not have the opportunity to observe. *(NASA/JPL/SSI)*

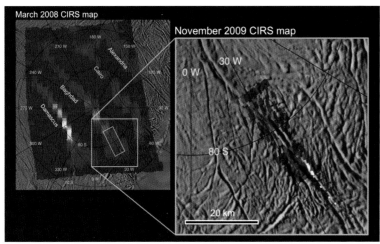

March 2008 CIRS map

November 2009 CIRS map

20 km

LEFT Heat radiation seen by CIRS emanating from the narrow fissure, Baghdad Sulcus on Enceladus. This image was taken during the flyby on 21 November 2009, and it is ten times sharper than the wider view showing the four 'tiger stripes' obtained a year earlier. The thermal image shows variations of temperature along the fissure, reaching 180K, 100° warmer than the background surface of the bright chilly moon. The new map shows a region (green box on the old map) between 5km and 10km wide, with the smallest features on the thermal map less than 1km across. *(NASA/GSFC/JPL)*

BELOW INMS measurements of Enceladus's plume vapour. In many respects, the plume composition (coloured bars) resembles that of comets (white bars). The presence of organics in liquid water makes Enceladus of major astrobiological interest. *(NASA/JPL/Southwest Research Institute)*

BELOW Saturn's aurora seen in the extreme ultraviolet by Cassini's UVIS instrument. The aurora was also observable using the Hubble Space Telescope. The white box at the upper left indicates a localised emission that tracks with Enceladus's orbital position, indicating that this emission is due to material coming from Enceladus's plumes. *(NASA-Caltech/U. of Colorado)*

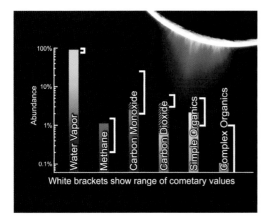

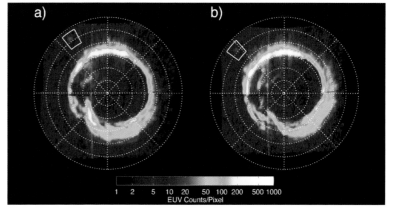

Mars

Venus

Earth and moon

Cassini's observations have brought a wider perspective, opening our eyes to new worlds and new possibilities.

Ground segment

The Deep Space Network not only transmits commands to Cassini and receives spacecraft data, it also performs tracking via Doppler and ranging measurements to precisely measure the vehicle's trajectory and thereby allow the planning of correction manoeuvres. The DSN antennas are also used for Cassini radio science, measuring the trajectory during satellite flybys to estimate their gravity fields and hence infer details of their internal structures, measuring the refraction in the atmospheres of Saturn or Titan, diffraction in Saturn's rings, or bistatic reflection off Titan's surface.

The DSN stations are distributed around the globe so that as Earth rotates at least one station will be in view of a celestial target. The main station is at Goldstone in California's Mojave desert, several hours' drive from JPL where NASA deep space missions are controlled. The remote dry desert conditions help to reduce the incidence of rainstorms which can interfere with radio communication and minimise radio noise from human activity. Another station is near Canberra, Australia, and the third near Madrid, Spain. These three main stations are separated by roughly 120° of longitude (or 8hr of local time) and together they maintain continuous coverage of planetary targets. Depending on the season and position,

ABOVE A magnificent ISS mosaic (using some 141 WAC images in red, blue and green filters) of a backlit Saturn and its ring system, including the diffuse E-ring as a halo. The Earth and Moon are also visible. This remarkable montage, some 650,000km across, was taken on 19 July 2013, when Cassini was in Saturn's shadow. *(NASA/JPL-Caltech/Space Science Institute)*

a target may be visible from two stations, and for high priority events both might be operated simultaneously to achieve a few hours of overlap rather than handing over from one to the other.

Each of the main DSN stations has a single 70m dish and several smaller ones (34m, 23m and smaller). The 70m dishes offer the highest performance, allowing Cassini to send its data at the highest rate, but these assets are heavily oversubscribed – especially when Mars, buzzing with spacecraft, is in the same part of the sky as Saturn. A 34m dish can receive data at only about 25% of the rate of a 70m dish, so an 8hr pass will only partly empty Cassini's

BELOW A fine view of Saturn's north pole in summer (obtained in April 2016) showing the polar hexagon and some slivers of ring-shadow on Saturn. The curved shadow of Saturn on the rings is just as Christiaan Huygens imagined it would be in the mid-17th century. *(NASA/JPL-Caltech/Space Science Institute)*

data recorders. But it is possible to connect antennas together by a technique called 'arraying' in order to synthesise the capability of a larger antenna, so this is sometimes done to achieve a capacity close to that of a 70m dish.

The size of these antennas is matched by exquisite sensitivity. The receivers onto which the dishes focus the radio energy use masers (microwave lasers) made of synthetic ruby, cooled down to 4.5K with liquid helium to minimise noise levels.

Often taken for granted, this impressive

and expensive infrastructure is essential to the operation of deep space missions. For Cassini, its reliability is particularly important since, unlike many Mars missions or space telescopes, the observing opportunities available to Cassini were fleeting. The relentless series of satellite encounters and other events required it to make an observation, then hose the data down to Earth to empty its recorders in time to begin a new observation. Apart from a few extra-high-value observations like close Titan encounters, which were sometimes deliberately played back twice for redundancy, there was usually only one chance to transfer the data, and Cassini had to proceed on the assumption that this would be safely received on Earth.

As an example, on 9 August 2009, the 70m antenna at Goldstone (originally a 64m dish that was built to support the Mariner 4 flyby of Mars in 1966) was tracking the Mars Reconnaissance Orbiter prior to the start of a Cassini pass when the massive antenna's azimuth bearing seized. Although a smaller antenna was scrambled into operation (achieved by 'stealing' it from the New Horizons and Kepler missions – there is always a cost) to support the tracking and commanding planned for the pass, it was unable to recover the science data from Cassini's T60 pass over Ontario Lacus at Titan's south pole.

Although this and similar data losses represent less than 1% of Cassini's total data return, they are heart breaking when they occur.

Even the dual playback strategy is no guarantee. Many telemetry frames in the downlink to the Madrid 70m antenna after the T64 flyby were corrupted because storms there degraded the signal. The backup playback 24 hours later was also to Madrid, where the continuing bad weather impaired reception, so some data gaps existed even when the two downlinks were combined. Rain in Spain.

Ageing instruments

Assembly of the hardware for Cassini's instruments began in about 1994, some three years ahead of launch and a decade before the spacecraft arrived at Saturn. Although the vehicle systems were designed with redundant elements wherever possible – a factor that undoubtedly contributed to

the ability to maintain operations in space for two decades; e.g. the switch over to the B-thrusters, and the management of the RWAs – the science instruments and the observations that they make are typically not as robust.

Literally millions of spacecraft and instrument commands have been generated for the mission. Every image that Cassini has taken required pointing to be defined, the filters selected and the exposure time specified. In addition, the compression scheme (if any) and/ or binning of the data was selected. All of these details are documented on the Planetary Data System (PDS) where the mission results are made publicly available. Each scientific data file comes with a 'label' containing relevant geometric and other information. Unfortunately, just occasionally, a parameter was specified in error, or a time was chosen incorrectly. One major observation that was to have been made during Saturn Orbit Insertion was lost due to a commanding error that may have been related to 2004 being a leap year!

Often these suboptimal observations are not so much commanding errors as overwhelming challenges thrown by the Saturnian system at instruments not designed with such surprises in mind. For example, some of the radar altimeter echoes from Titan's seas saturated the instrument because the perfectly flat sea surface acted like a near-perfect mirror, dazzling the radar receiver. The radar receiver had an auto-gain function to adjust its sensitivity in real time, but this took several seconds to step the gain down to compensate. However, sometimes as Cassini swept across Titan's irregular coastlines the specular echo blinked on and off too rapidly for the auto-gain to track the changes in echo strength – which could vary by a factor of about a thousand.

Other times, observations were degraded by the sheer sensitivity of the instrumentation. CIRS utilised an interferometer with an oscillating optical bench to generate infrared spectra, and so periodic disturbances could appear as spurious lines in the spectrum. Some of these influences (fortunately reasonably easy to screen out by filtering in analysis) were related to the digital electronics; e.g. the 8Hz Real-Time Interrupt (RTI) clock on the spacecraft.

Similarly, the raw ISS images showed a faint

banding corresponding (via beating with the rate at which pixels were read out from the CCD chip) to a 2.1Hz fluctuation in a bias voltage. This issue wasn't seen on Earth. A terrestrial laboratory with air conditioning, computers, mains power and so on, is one of the electrically noisiest places in the solar system, and so it is often impossible to operate instruments at the exquisite sensitivity that they can achieve in the cold quiet of space.

Sometimes, sadly, the instruments worked perfectly but data were lost altogether, such as the occasional flybys where the pointer in the Solid State Recorder was incorrectly set, or where the DSN was unable to receive data that could only be transmitted once.

All in all, however, such losses were modest, perhaps amounting to (at worst) a couple of per cent of the total. This was in stark contrast to Galileo, where the entire scientific return from many precious flybys were lost because the spacecraft entered 'safe mode' in response to some spurious signal caused by radiation or electrical transients on its slip rings (which passed signals between the rotating and the stabilised sections of the vehicle). Cassini managers have often remarked how free of crises their operations were in comparison to other missions.

LEMMS, part of the MIMI instrument that measured low energy plasma, had a 15° field of view which was swept around in the X-7 axis plane of the spacecraft via a Y-axis stepper motor. In fact, when this actuator was operated, CIRS data were often degraded, probably due to mechanical vibration rather than electrical noise. Certain RWA wheel speeds caused similar vibration problems, with the interferometer stage on CIRS acting as a

ABOVE Workers pour a new oil-resistant epoxy grout beneath the azimuth bearing of the 70m Goldstone antenna during its first and only bearing replacement in 2009. Working at night avoided the worst of the desert heat. Occasionally the scientific exploration of distant worlds by sophisticated robots depends on such quotidian operations. *(NASA)*

sensitive seismometer! However, this ceased to be a problem for CIRS on 1 February 2005 when the LEMMS spin actuator failed, leaving the particle telescopes 77° off the Z axis. The particle telescopes, while no longer able to sample the full angular range, nevertheless continued to provide data in this fixed direction.

After 2005 day-of-year 248, the large element of the CDA High-Rate Detector developed a higher noise background that was suspected of being due to a relatively large impact on the detector. As a result, the detection threshold had to be increased, lowering the unit's effective sensitivity slightly. Perhaps rather than a failure, this might be legitimately considered 'wear and tear'. The quality of Cassini measurements was reliant on careful operation and vigilant calibration to monitor such changes (e.g. slight wavelength shifts in the VIMS instrument) and ensure they could be taken into account during the ensuing scientific analysis.

The CAPS instrument was operated extensively during the cruise. These several years of operation saw only some minor software issues, related in part to a 700µsec timing difference in run-time between the engineering model instrument on which the software had been tested and the flight instrument. Software updates and reboots fixed the problems. But a few months after SOI, while the actuator motor to sweep around the instrument's particle aperture continued to operate as intended, the sensor which reported the position of the aperture flat-lined. Engineers could still determine its position sometimes (by driving the unit into one of its end stops!) but thereafter the control and estimation of the sensor geometry wasn't as precise as intended.

The Helium Vector Magnetometer degraded over time, possibly due to a gradual loss of the helium gas, and was inoperable by late 2005, but the Flux Gate Magnetometer was able to address all of the relevant science objectives of the mission with only a modest science loss.

The Ka-band translator of the Radio Science Subsystem failed during the interplanetary cruise, after the first gravitational wave experiment. Tracking measurements for gravity field determinations would have been four times better at Saturn if this had still been in operation. The ultra stable oscillator of the system appeared to fail in December 2011, and after tests in July 2012 indicated a hard failure it was decided to power the unit down. Fortunately this was well after the nominal mission. In many cases (such as Titan gravity measurements) data could be obtained instead by establishing a coherent two-way link with the ground. The ultra-stable oscillator failure also caused some degradation of subsequent radio occultations since a few seconds of data were lost when the link was re-acquired during egress (when the spacecraft emerged from behind Titan or Saturn).

CAPS began to suffer some short-circuit issues in June 2011. It seems that tin 'whiskers' had grown from solder joints and upon contacting another conductor had drawn current. These problems recurred in June 2012, causing the Solid State Power Switches to trip. Eventually, perhaps unfairly, concerns about failure propagation beyond CAPS led to the decision to leave the instrument switched off.

Even a 'Flagship' mission such as Cassini wasn't immune to failure, but largely because of the resilience afforded by redundancy the instrumentation and systems continued to produce high-impact science long after the spacecraft's warranty had expired.

Planetary protection

All signatories to the United Nations 1967 Outer Space Treaty (the *Treaty on Principles Governing the Activities of States in the Exploration and Use of Outer Space, including the Moon and Other Celestial Bodies*) are obliged to comply with a number of internationally agreed policies on spacefaring to ensure that no party shall cause potentially harmful interference with activities in the peaceful exploration of space.

Article IX stipulates: 'Parties to the Treaty shall pursue studies of outer space, including the Moon and other celestial bodies, and conduct exploration of them so as to avoid their harmful contamination and also adverse changes in the environment of Earth resulting from the introduction of extraterrestrial matter and, where necessary, shall adopt appropriate measures for this purpose...'

This treaty has been embellished by the Committee on Space Research (COSPAR) to enact standards to be followed in missions to different categories of planetary target. The COSPAR policies guide NASA's efforts via US government directive NPR 8020.12 entitled: *Planetary Protection Provisions for Robotic Extraterrestrial Missions*. NASA has a senior scientist, the Planetary Protection Officer, who is charged with ensuring that missions comply with this policy; a job that takes significant determination, since the requirements are sometimes onerous except for 'Category I' targets which are sterile, inhospitable worlds like Mercury and the Moon.

Category II is: 'Any mission to locations of significant interest for chemical evolution and the origin of life, but only a remote chance that spacecraft-borne contamination could compromise investigations.' It requires only that the spacecraft and its impact site be documented. Titan falls into this category. While battery-powered Huygens had no significant requirements (with the meagre heat from the 20-odd RHUs inside, the probe's interior would have cooled to just a couple of degrees above the bitter 94K ambient temperature, within days of the mission's end), proposals for future missions like TiME (see next chapter) did have to show that even if the plutonium-fuelled GPHS bricks fell out of its small power source they wouldn't melt their way into Titan's interior.

But once the plumes of Enceladus revealed the existence of a near-subsurface liquid water reservoir on that moon, this discovery raised the planetary protection stakes in the Saturnian system, prompting a US National Research Council study on Planetary Protection for Icy Moons. Because in principle Enceladus has environments in which terrestrial microbes could survive if they were introduced, the disposal of the Cassini spacecraft became a planetary protection issue.

Enceladus (like Mars or the Jovian moon Europa) is classified as a Category III or Category IV target, depending on the type of mission.

Category III applies to flyby and orbiter missions to places of significant interest in terms of chemical evolution and/or the origin of life, posing a significant chance that contamination could compromise investigations. The requirements include extensive documentation (including an inventory of organic materials) and possibly trajectory biasing (to lower collision probability), clean room assembly and possibly bioburden reduction.

Category IV addresses landers or probe missions to the same locations as Category III. The measures to be applied depend on the target body and the planned operations. Sterilisation of the entire spacecraft may be required for landers and rovers with life detection experiments, and for those that land in or move to a region where terrestrial microorganisms may survive and grow, or where indigenous life may be present. For other landers and rovers, the requirements would be for decontamination and partial sterilisation of the landed hardware.

Some of the aggressive sterilisation measures that were applied to the Ranger moonshots in the 1960s are believed to have stressed their components such as to make them prone to failure. It is estimated that the planetary protection efforts on the Ranger capsule accounted for some 3–10% of the total program cost. On a modern Mars mission, sterilisation (and selecting and screening components which can survive the sterilisation process) might cost in the order of $100 million; a substantial fraction of the total mission cost. If these requirements were imposed on an Enceladus mission, would it still be affordable?

Planetary protection issues were a major factor in the decision to end the Cassini mission by plunging it into Saturn in order to eliminate the possibility of the 'retired' spacecraft colliding with and potentially contaminating Enceladus. The same logic, in part, had driven Galileo's disposal into Jupiter to avoid impact with Europa.

BELOW Installing the RTGs on Cassini through a hatch in the fairing. The radioisotope power source, while vital for Cassini's technical feasibility, added significant complications in launch approval and safety, and later in planetary protection. The suits, masks, and gloves are standard for clean assembly of spacecraft to avoid contamination of delicate mechanisms and optical surfaces etc., and also help to prevent a large build-up of microorganisms during this work. *(NASA)*

Chapter Nine

Cassini and beyond

Cassini's last, busy orbits are planned for summer 2017 before the spacecraft will burn up in Saturn's atmosphere. But its data will live on, feeding scientific enquiry for decades to come. And just as Cassini originated from the questions raised by Voyager a decade and a half before Cassini reached the launch pad, imaginative concepts for new missions to Titan and the Saturnian system are now reaching the point where they may soon go ahead.

OPPOSITE An artist's impression of the Titan Mare Explorer, the most detailed proposal of a post-Cassini mission so far. Simulation tools were used to estimate the vehicle's dynamic response to the waves predicted for 2023 on Titan's second-largest sea, Ligeia Mare.
(Johns Hopkins Applied Physics Laboratory/Lockheed Martin)

The Grand Finale

All good things come to an end, and after 20 years in space, Cassini and its Radioisotope Thermoelectric Generators are set to be safely disposed of in Saturn's near-bottomless atmosphere. Getting to this point has required careful husbanding of its fuel, as well as management of other consumables such as RWA revolutions. At the time of writing (late 2016), Cassini was down to its final 1% or so of propellant, a fraction so small that it was difficult to gauge accurately.

But Cassini will continue to produce scientific data to the last, taking images until a few hours before it plunges into Saturn, and then bravely transmitting in-situ data as its thrusters struggle to fight the ever-growing atmospheric drag torques just hundreds of kilometres above the planet's cloudtops. Eventually, these will win and the spacecraft will tumble.

Scientists may use telescopes on Earth to attempt to observe the Cassini entry, like a giant meteor. (An attempt to detect Huygens blazing a trail through Titan's atmosphere in 2005 was unsuccessful, in part owing to bad weather on Mauna Kea in Hawaii and the failure of a key instrument on the Hubble Space Telescope.)

Even amateur observers equipped with 30cm telescopes have detected the entry of several meteoroids of ~100 tons at Jupiter. So, knowing the place and time of entry should enable large observatories to catch sight of Cassini's fiery demise. The emission spectrum of the glowing air, tortured by the spacecraft's hypersonic passage at 30km/s, might yield secrets about Saturn even as the vehicle burns up.

But there is much to be done before this Viking funeral.

The final 'targeted' Titan passage of the Solstice Mission (T126) will nudge Cassini's orbit such that the periapsis hops from being just outside the F-ring to just inside the D-ring. This elliptical orbit will offer several long-range Titan observations to document the changing cloud patterns as late as possible in the seasonal cycle – north polar clouds, mirroring those seen in the south upon Cassini's arrival in 2004, are expected to intensify in 2017.

However, the main new opportunity will be to study Saturn and its rings up-close. Like Juno at Jupiter, remote sensing will reveal exciting new details of the planet, and the low periapsis will expose higher harmonics of its gravity and magnetic fields to scrutiny. The first active radar scanning of the rings will be attempted, and the Doppler shift of the echo should yield new insight into the dispersion of velocity amongst the ring particles. Cassini's other instruments will study the dust and plasma composition in an as-yet unexplored region. So a whole new set of discovery possibilities may be opened up.

But these final orbits pose new challenges to mission planners. During the periapsis passages, the planet and its rings will occupy a large part of the sky, so the star trackers will have to be disabled and Cassini's attitude controlled solely by the inertial reference units. Furthermore, because Saturn is warm compared to the depths of space, and its proximity will leave nowhere dark for Cassini to point its instrument radiators when making science observations, the VIMS and CIRS observations will be degraded for a number of hours after close approach because their sensitive detectors will have warmed up.

Charged particles trapped in Saturn's magnetic field can cause errors in computer memories (as well as little bright streaks in CCD images). The solid state data recorders on Cassini can mop up bit-flip errors caused by particles lancing through their silicon memory chips if they occur slowly enough. On each recorder of 2Gbits, several dozen bits/hr are typically corrupted, detected and repaired. Sometimes, a couple of bits in a single word were flipped. These errors could be detected but not fixed. Such one-in-a-billion errors were tolerable in science data. A special partition in memory was used for flight software that was more sensitive.

But occasionally there have been been bursts when thousands of single-bit errors and several dozen double-bit errors occurred per hour. These episodes were shortly before and after close periapsis passages, when Cassini was passing through Saturn's radiation belts (analogous to Earth's Van Allen belts, where the magnetic field traps charged particles captured from the solar wind). These belts are much less intense at Saturn than at fearsome Jupiter because Saturn's magnetic field is weaker, and the rings intercept many of the trapped particles, but the belts may cause some disruption as Cassini makes frequent close periapsis passages in the proximal orbit phase. Vigilance may be demanded in these final months of intense operation in a somewhat new environment.

Cassini's legacy

Once Cassini's entry has turned an exquisite artefact of technology into particles of smoke and perhaps a few blobs of molten metal which descend slowly towards Saturn's core, the enterprise will still not quite be over. After a year or so of data analysis and closeout documentation, the project will be wound up as the accountants close the books.

Many senior scientists, engineers and managers, having spent their best years on this project, may take this watershed moment as an opportunity to retire. Their younger colleagues will pursue work on other projects – indeed most will have already done so, sometimes 'moonlighting' on Cassini for only a small fraction of their time for the past few years. Hopefully the hard-won skills, tools, and close professional relationships acquired on Cassini will be usefully applied in other projects.

Office and laboratory space for operations planning, calibration and testing that have been securely held for decades by the exigencies of a major flight project, will go up for grabs in the turf wars that pervade organisations such as universities. Shelves of meeting minutes, test reports, software manuals and other minutiae will be bundled off to boxes in the basement archive, or perhaps shredded. Engineering model hardware and flight testbeds that were carefully maintained for all these years will be junked, with the accountants duly noting down the tracking numbers of the little property tags required for government-procured equipment. Some memorable items will find their way into museums, as this author's contribution has done. Many others are sure to sit in a proud scientist's office like a war trophy, generally unappreciated until some occasional curious visitor asks the provenance of a strange-looking metal box on a shelf.

But the enduring legacy of the Cassini-Huygens mission will be the data that was sent back, and the use to which it is put. Thousands of scientific papers were written, and over 200 doctoral theses were penned on Cassini's instruments and what they discovered. The cliché is true: the textbooks are being rewritten.

In fact, the science will continue long after the end of the project. Even before Cassini's arrival gave new context for comparison with Voyager's observations 24 years earlier, Voyager data were being analysed with new methods in 2003. Cassini's voluminous return from the phenomenologically rich Saturn system will surely be fodder for new enquiries for decades to come. This exploitation will require the data to be archived openly. This process once took place in a handful of repositories worldwide, which stored data tapes and photographic prints that researchers needed to physically visit to examine.

The growth of the World Wide Web – which did not exist when the Cassini project began – now makes the scientific data accessible by anyone, from anywhere. But to do so you first have to know it exists, to find it, and to know enough about how it was obtained in order to make sense of it.

Data are held electronically by NASA's Planetary Data System, in one of several 'nodes' (organised by discipline; for example, the 'Atmospheres' Node at New Mexico State University in Las Cruces). ESA, adopting the same standards as the PDS, now has its own Planetary Science Archive. These standards, hammered out in long meetings and described (of course) in voluminous documents such as

BELOW Cassini's 'F-PROX', F-ring and proximal orbit phase in 2017, sees the periapsis duck inside the F-ring after Titan flyby T125.
Then the final Titan encounter T126 moves the periapsis inside the entire ring system in order to make a series of very close passes just a few thousand kilometres above Saturn's cloudtops, prior to its disposal in the atmosphere. *(NASA/JPL)*

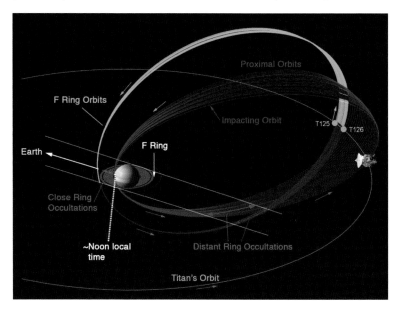

Software Interface Specifications (SIS), ensure that data can be read and understood, ideally without proprietary software.

Data files must be supported with label files that describe the format – which column of numbers means what, and the units the numbers are expressed in. The pedigree of derived products must be traceable. Is the wavelength list for a VIMS spectrum that was thought to be the right one determined after launch or is it corrected for a small calibration drift that was noticed after 2010? These steps in the chain have to be documented.

The Huygens dataset was archived in more or less one go in 2006. Cassini's continuing results have been archived in roughly three-monthly chunks in order to balance the workload in each delivery against the need to be prompt. The archival process – now a much more formal obligation in NASA projects than it once was – is very demanding. External peer reviewers check the archive before public release, to verify that necessary information is present, that file formats are correct and so on, and data preparers must resolve any issues before the process is declared complete.

But with this often tedious investment of effort, the data that was sent to Earth by Cassini, the ultimate purpose of its painstaking design, construction and operation, can be of lasting value to humanity. And not just science; there is a cultural impact too. Snippets of sound – the rush of air and buzz of the radar altimeter recorded by Huygens, or the insonified whistles and chorus of Saturnian radio emissions that

were recorded by RPWS – have made their way into music. And Cassini's numerous images have been assembled into haunting movies and the places Cassini has explored (notably Titan) have become the setting for fiction.

After Cassini

Ideas for the exploration of Saturn and Titan beyond Voyager have swirled around since the mid-1970s, and even then (before the thickness of its atmosphere was measured) the importance of that atmosphere for future missions was recognised.

One idea was 'aerocapture' wherein a spacecraft in a slender biconic or 'bent-cone' shaped heatshield could dip into Titan's atmosphere in order to efficiently slow down from its interplanetary trajectory, but then climb out into orbit around Titan or Saturn. This approach meant buttoning up the spacecraft in a heatshield, but would avoid the massive fuel tanks and engine burn that Cassini needed for orbit insertion. French and American scientists also advocated the possibility of balloon exploration. But all these ideas fell into abeyance for 15 years, once the development of the Cassini-Huygens mission began.

In the late 1990s, however, with Cassini safely on its way, serious thinking began again. The scientific priorities for post-Cassini exploration at Saturn could only be guessed at, but NASA committees judged that Titan's surface chemistry – which would surely be complicated but Cassini-Huygens was not well-equipped to measure – would likely be top of the list. And of course there was interest in a Saturn probe (as conceived in the SO2P concept that was a forerunner of Cassini) as well as in further, more detailed inspections of the rings.

A workshop in Houston, Texas, in February 2001 identified a number of concepts for Titan, including a rather impractical 'aerover' incorporating three large spherical tyres that could be inflated with helium to permit balloon traverses between periods of roving the surface. While Mars exploration in the early 21st century became dominated by rovers, the uncertain trafficability of Titan's surface discouraged wheeled vehicle exploration there, and instead attention focused on aerial mobility

BELOW Cassini data are archived on the PDS for all to use. Preparing, checking, and providing a search capability for the archive is a major undertaking. *(NASA)*

– balloons or airships – which could exploit the thick atmosphere.

A challenge for these platforms is how to access surface material for analysis without requiring the craft to land. Ideas that have been kicked around (and even prototyped for laboratory tests) include a tethered coring penetrator, rather like a hollow harpoon that would be winched back up to a hovering airship, or a 'touch and go' sampler using abrasive contra-rotating wire brushes that could be dangled down a tether.

Around this time, the potential for heavier-than-air flight at Titan was recognised because not only does the thick atmosphere help, as it does for balloons, but the reduced gravity makes flying an aircraft easier. Aeroplanes, helicopters and even tilt-rotor aircraft were proposed. In particular, it was noted that a given helicopter configuration could hover on Titan with 38 times less power than is needed on Earth. Whilst such a vehicle could not fly continuously given the low power-to-weight ratio of RTGs, it could trickle-charge a large battery over a Titan night lasting eight Earth days and then fly for several hours in daylight.

Around 2002, NASA explored the development of much higher-powered systems – nuclear reactors for space. In particular in a program called Prometheus, it advocated a Jupiter Icy Moons Orbiter (JIMO) to follow up the findings from Galileo at Europa. It was even imagined that future versions could be tasked to Saturn, but this giant project was shelved as soon as its cost proved to be in the order of tens of billions of dollars.

Such a large nuclear-electric spacecraft was portrayed as the delivery system for a Titan airship that was studied by NASA for its Revolutionary AeroSpace Concepts (RASC) program. This imagined an airship performing surface analysis using a PAARV (Planetary Autonomous Amphibious Robotic Vehicle) – a small tracked vehicle with a variable-buoyancy ballast tank that would enable it to operate in a liquid as well as on dry land. This vehicle would be lowered on a tether while the airship hovered above a site of interest.

Around this time, significant effort was devoted by NASA's In-Space Propulsion program to considering future technologies, and

LEFT A futuristic helicopter inspecting the Huygens landing site, as imagined circa 2001. The co-axial rotors not only provide efficient packaging inside an entry shell, but also avoid the need for a tail rotor. A rotorcraft provides the ideal combination of long-range mobility and surface access for exploring Titan's diverse terrain. *(James Garry)*

BELOW Top left: A Titan entry capsule being released from a JIMO-like nuclear-electric interplanetary spacecraft (the flat surfaces are thermal radiators, rather than solar panels). Top right: An airship deploying into the atmosphere of Titan from the initial vertical orientation under a parachute. Bottom left: PAARV being lowered from the airship on a tether. Bottom right: PAARV driving across the floor of a hydrocarbon lake. *(Excerpts from a NASA animation)*

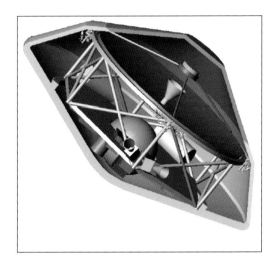

RIGHT A future Titan orbiter could be more massive and capable if it was delivered by aerocapture, rather than by braking into orbit using rocket propulsion. For simple Titan orbit insertion, high lift-to-drag isn't required, and a conventional entry shell design will suffice, as with Huygens and Mars probes. *(NASA)*

ABOVE A 13m airship envelope under test at cryogenic temperatures – note the silvered walls of the test 'tent'. *(J. Hall/M. Pauken/JPL*

RIGHT Engineers test a satellite communications antenna on an airship out on a desert dry lake bed to evaluate how well it can maintain its pointing accuracy while the vehicle swings in the wind. *(J. Hall/JPL)*

in particular the use of aerocapture. As in the 1980s, Titan was recognised as an appealing target for aerocapture because the technique is easier for this moon than in the atmospheres of the giant planets, and the mass delivery performance is attractive.

After the exciting view of Titan provided by the Huygens probe, the hot air balloon or 'Montgolfière' became a preferred concept (named after the Montgolfier brothers who made the first flight in such a balloon). It would be simpler than an airship, and not susceptible to leaks of helium through its large envelope. It would have to be packed tightly in an aeroshell for years on its way to Titan and then, on being deployed, would be cold and stiff. The waste heat from an RTG could be employed to provide buoyancy. A Montgolfière would lack the horizontal control authority of a buoyant gas airship, but it could easily manoeuvre up and down by modulating the heat fed into the envelope or by venting hot air through a crown valve. As with recreational hot air ballooning on Earth, some horizontal control may be able to be achieved

BELOW An artist's concept of a Montgolfière (hot air) balloon over the shore of one of Titan's seas. *(T. Balint/NASA)*

by changing the altitude to a level where the winds are close to the desired direction.

The breadth of Titan science and interest in Enceladus prompted NASA to commission 'Flagship' (i.e. Cassini-class) mission studies for these targets. These funded studies achieved a much higher level of technical detail than the previous suggestions.

The 2007 Titan Explorer study led by the Johns Hopkins Applied Physics Laboratory, noted the wide range of scientific investigations. A variety of architectures were considered but soon reduced to a triple-platform mission (orbiter, lander and balloon), packaged on a single Atlas 551 launch vehicle. An important constraint was that the radioisotope power sources specified as being available for these new proposals were a 100W-class Multi-Mission Radioisotope Thermoelectric Generator (MMRTG) and an Advanced Stirling Radioisotope Generator (ASRG) which at that time were being developed by NASA. Such a unit would use a small reciprocating engine to convert heat from plutonium into electrical power with an efficiency about four times better than an RTG. (At the time the American Pu-238 inventory, which had even been supplemented in the 1990s by buying some material from Russia, was looking meagre so there was an incentive to introduce a more efficient converter that would need less plutonium for a given electrical output than an RTG.)

The orbiter would use aerocapture to efficiently brake into orbit around Titan. Although this would be a packaging challenge, it permitted the orbiter to have a formidable capability of some 1,800kg, including about 170kg of instrument payload and consumables for four years of Titan operations, including propellant to conduct dipping orbits for 'aerosampling'. From orbit, the craft would completely map Titan's surface at resolutions better than attained by Cassini and its radar altimeter could directly measure the tide in the hydrocarbon seas whose existence was discovered only towards the end of the study period.

In order to bridge the global and local scales sampled by the orbiter and lander, a Montgolfière balloon was included to perform a regional survey. A one-year lifetime was assumed, with a

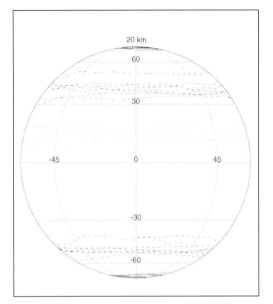

LEFT Simulated balloon trajectories on Titan. Each colour corresponds to a different starting latitude. In one year at an altitude of 10–20km, the balloon would make several circumnavigations in a latitude band of about 10–20°, surveying a wide range of terrain types. High-latitude balloons would tend to drift poleward. *(T. Tokano/author)*

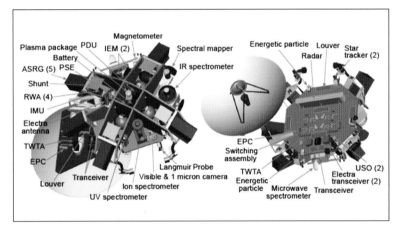

ABOVE The orbiter of the 2007 Titan Flagship study packed a dozen instruments and five ASRG generators in a compact configuration in order to fit inside a heatshield for aerocapture. *(NASA/ Johns Hopkins Applied Physics Laboratory)*

float altitude of about 10km, where zonal winds should permit the balloon to circumnavigate Titan one or two times. Since the 2kW of waste heat from the RTG would be fairly small, a double-walled balloon envelope would limit the heat loss to the cold ambient atmosphere.

With early Cassini data showing the diversity of Titan's surface, the question of a landing site had to be confronted, at least in a preliminary ('existence proof') way. The landing site was chosen conservatively as Belet, a large equatorial region covered in organic-rich sand dunes that provide a large, scientifically appealing target with feasible landing characteristics.

After arrival using an airbag system to limit the impact loads and Mars Pathfinder-like petal deployment, the lander would use a gimballed 0.5m X-band HGA to communicate with the orbiter; this antenna would also permit a modest Direct-to-Earth communication of

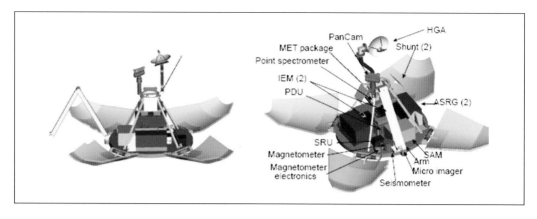

RIGHT The configuration of the lander for the 2007 Titan Flagship study. Unfolding the petals after landing on a field of dunes will ensure its upright orientation. A robot arm performs surface sampling and deploys a seismometer package onto the surface. A mast-mounted camera and spectrometer suite undertakes a local survey. A small High-Gain Antenna provides for direct-to-earth as well as communication with a relay orbiter. *(NASA/ Johns Hopkins Applied Physics Laboratory)*

200–300bps if necessary. Orbiter overflights for communication would occur in three-day groups every eight days, each overflight being separated by the orbital period of 5hrs.

The lander would be powered by two ASRGs and hence be able to operate almost indefinitely. Valuable synergetic measurements of the magnetic field by the orbiter and lander simultaneously would facilitate separation of the induced signature of a subsurface water ocean. A robot sampling arm would deploy the magnetometer and a seismometer onto the surface in order to minimise 'noise' from the lander. The arm would have a microscope camera, as well as illumination for spectral measurements (providing ground truth for the orbiter's spectral map, which must observe through the thick atmosphere). The panoramic

camera and point spectrometer would survey the landing site and support sampling operations; context being provided by images acquired by a descent imager.

The possibility of augmenting the lander science with a small lander-launched aeroplane was advocated in the 2007 study; by that time 'drones' or Unmanned Aerial Vehicles (UAV) were becoming commonplace tools in warfare on Earth.

A 1kg battery-powered UAV would simply fly off the side of the lander for a couple of hours. This idea was nicknamed the Titan Bumblebee owing to two analogies with that insect. First, the thrust-to-weight requirement for vertical take-off would allow the vehicle to hover, just like a bee. Second, a significant portion of the energy budget of a small UAV in the cold Titan environment must be expended simply to keep the vehicle warm, suggesting a parallel with the specific heat management adaptations acquired by the bumblebee for the subarctic environment to which it is evolved.

Another NASA Flagship study, this one intended for Enceladus, was carried out by the Goddard Space Flight Center. It imagined an orbiter to fly through the recently discovered plumes, and a small lander.

Meanwhile, an (unfunded) European study called TandEM (Titan and Enceladus Mission) assembled a coalition of scientists and imagined an orbiter with a large number of instruments which would perform several Enceladus flybys and deliver penetrators to its surface in advance of entering orbit around Titan, while a second spacecraft would deliver a Montgolfière and perhaps several landing probes. Although the proposal lacked detail, its innovations included a resonant trajectory used to visit Titan and Enceladus and the hot air balloon.

BELOW An Enceladus orbiter, powered by ASRGs, with dust impact shields (like Giotto's) for flying safely through the plumes. *(NASA/GSFC)*

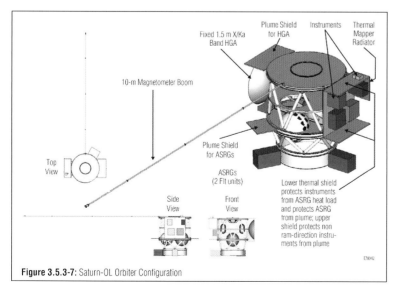

Figure 3.5.3-7: Saturn-OL Orbiter Configuration

The Montgolfière would have a radius of 6.5m, a mass of 28kg for a skin thickness of 30μm made of a film of Kapton coated with another polymer on both sides, fused into place. Unlike the Titan Flagship study, this TandEM balloon design called for a single-wall envelope. It improved its altitude stability in the temperature gradient of Titan's atmosphere by adding a small hydrogen balloon. Deployment would start at an altitude of around 30km, and the balloons inflated while descending to 3km, where the tanks and inflation system would be jettisoned. Then the vehicle would ascend to a nominal float altitude of 10km.

A new NASA study, Titan Saturn System Mission (TSSM) started in 2008. Led by JPL, this study was directed to consider a Titan mission that would also study Enceladus, with optional accommodation of in-situ elements to be examined in a coordinated parallel study by ESA. A further constraint ruled out using aerocapture for orbit insertion. This then required that the spacecraft use electric propulsion (ion thrusters, like JIMO, but powered by large solar panels that would be jettisoned as sunlight became too faint) in order to reach Saturn, and then chemical rockets to enter orbit around the planet. Following a resonant tour of Enceladus

flybys, it would burn into a high Titan orbit which would then be gradually lowered using aerobraking (a slower, gentler process than aerocapture, and now routine for Mars missions; the slower braking from a high orbit saves less fuel, but causes only modest drag and heating, hence no entry shell is necessary). A 'lake lander' and a Montgolfière would be deployed during the resonant tour phase. Planetary alignments meant that a launch in September 2020 would not be able to benefit from a Jupiter flyby, and therefore the mission would not arrive at Saturn until October 2029.

The battery-powered lake lander would be released by the orbiter on its second Titan flyby, targeted to Kraken Mare, a polar sea at a latitude of about 72°N. It would make atmospheric measurements during parachute descent, sampling the polar atmosphere which may have interesting differences from the low-latitude profile acquired by Huygens. The lander would splash into the hydrocarbon sea. Measurements on the sea surface would include the chemical composition of the liquid itself, dynamics of waves, and the depth of the sea. Because the mission would occur in northern winter the only ambient illumination would be diffuse twilight, therefore a surface science lamp

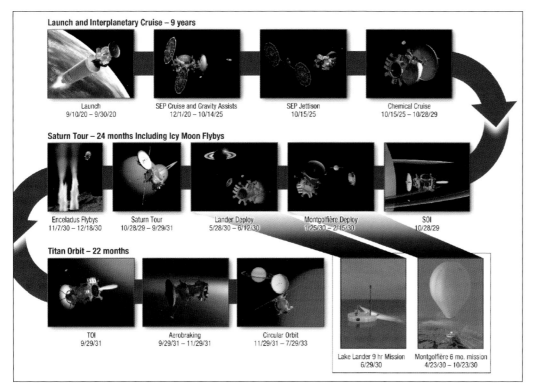

Launch and Interplanetary Cruise – 9 years

| Launch 9/10/20 – 9/30/20 | SEP Cruise and Gravity Assists 12/1/20 – 10/14/25 | SEP Jettison 10/15/25 | Chemical Cruise 10/15/25 – 10/28/29 |

Saturn Tour – 24 months Including Icy Moon Flybys

| Enceladus Flybys 11/7/30 – 12/18/30 | Saturn Tour 10/28/29 – 9/29/31 | Lander Deploy 5/28/30 – 6/12/30 | Montgolfière Deploy 1/25/30 – 2/15/30 | SOI 10/28/29 |

Titan Orbit – 22 months

| TOI 9/29/31 | Aerobraking 9/29/31 – 11/29/31 | Circular Orbit 11/29/31 – 7/29/33 | Lake Lander 9 hr Mission 6/29/30 | Montgolfière 6 mo. mission 4/23/30 – 10/23/30 |

LEFT A schematic timeline of the TSSM mission. The need to accommodate Enceladus and Titan science without employing aerocapture entailed an elaborate mission profile. A solar-electric propulsion cruise stage that deployed large panels would be discarded en route to Saturn. *(NASA/JPL)*

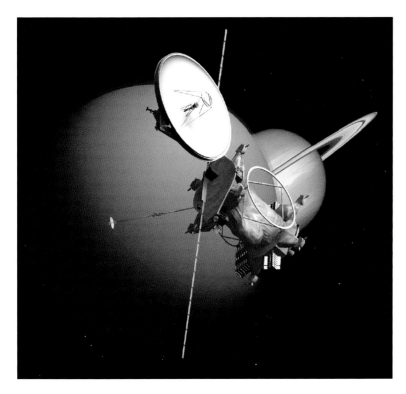

ABOVE An artist's impression of the TSSM orbiter with long ground-penetrating radar antennas and High-Gain Antenna deployed at Titan. The rings that carried the ESA Montgolfière and the lake lander (both released prior to orbit insertion) are visible. ASRGs with cooling fins can be seen at the base. (NASA/JPL)

to contract the orbit by aerobraking and to perform direct sampling of the high-molecular-weight photochemical products that are manufactured in the upper atmosphere. After circularisation, the orbiter would undertake a mapping task similar to that envisaged in the 2007 study, although in this case the TSSM orbiter could accommodate only a smaller payload and a two-year orbital mission.

By the end of the decade the prospects for a Flagship-class mission to Saturn had receded, with Europa and Mars being deemed by a US 'Decadal Survey' to be the priorities for that class.

Some important and interesting Montgolfière developments continued in cooperation with the French space agency CNES, simulating heat flow from an RTG to the interior of the balloon, measuring heat transfer in double-wall balloons at low temperature, and testing inflation during descent with air-drops. But after a couple of years, this work was discontinued.

The Decadal Survey did recommend a Saturn probe as a ~$1 billion class 'New Frontiers' mission that would complement the Galileo probe at Jupiter. But such a mission, which would make measurements for only an hour or two, has relatively narrow scientific appeal. The rings also pose a mission design challenge, since the danger of dust impact means a relay spacecraft must avoid them, and their material may block the radio signal for some geometries.

Nevertheless, interest in Titan and Enceladus persisted with a move to small standalone missions even though these would lack the

would be needed for imaging. This phase of the mission was to last only a few hours.

After the Saturn Tour Phase, the orbiter would brake propulsively into an elliptical orbit around Titan with a periapsis low enough

RIGHT A small model balloon attempts to inflate as it descends, as it would on Titan. In this test it was dropped from a manned hot air balloon and photographed from the ground. The test demonstrated that the opening in this instance was too small to permit rapid inflation, and the airflow pushes in the side of the balloon. (CNES)

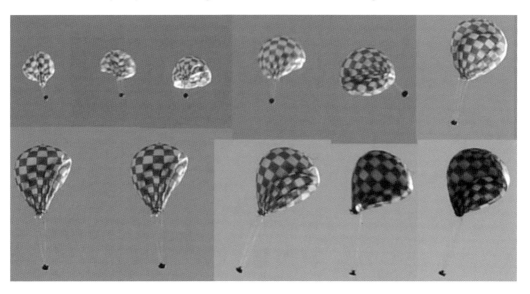

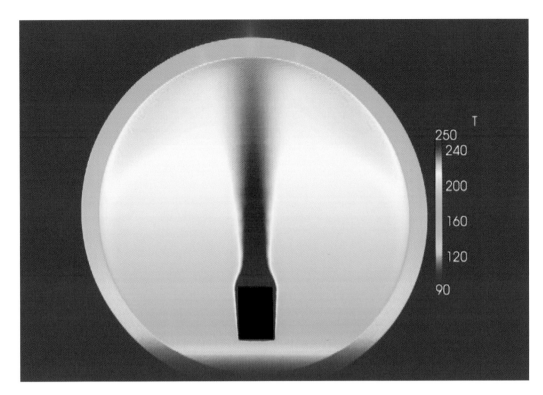

LEFT Computational fluid dynamics simulations of the heat transfer challenges facing a Montgolfière in Titan's atmosphere. Colours are the simulated temperatures, with the warm red plume rising from the RTG (black). This simulation shows how the double wall limits the heat loss. Note that the temperature is not uniform either in the inner envelope nor between the walls. *(T. Colonius/Y.Feldman)*

RIGHT A super-pressure balloon prototype (in this case an aluminised envelope intended for Venus). A similar balloon was imagined for the Titan Aerial Explorer (TAE) proposal. *(NASA/JPL)*

BELOW An imagined Saturn probe that is similar to the Galileo probe that penetrated the atmosphere of Jupiter. Note the ring shadow – the presence of the rings somewhat complicate a Saturn probe's data relay options. *(NASA)*

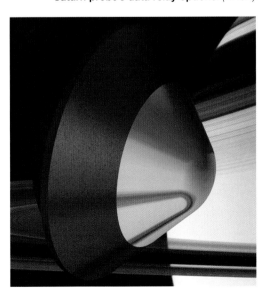

RIGHT AVIATR would be delivered to Titan inside an aeroshell with a carrier stage that resembled those which delivered Mars rovers. As with Cassini, the radioisotope power sources would need to be installed at a late stage in integration, and so a dedicated hatch in the aeroshell was included to allow access on the launch pad. (J. Barnes)

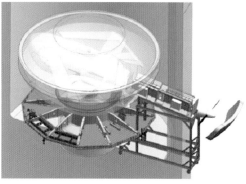

synergies between platforms which in-situ plus orbiter missions provide.

A super-pressure balloon concept was proposed to ESA's M3 mission call for a Medium-class mission that would cost roughly $500 million, similar to NASA's Discovery program of small missions. This was the 170kg Titan Aerial Explorer (TAE). Although a helium balloon is vulnerable to leaks, and the mass of the tanks to hold the 28kg of gas makes such a platform less efficient in some ways than a Montgolfière, the inflation scheme is simpler (and indeed was demonstrated on two Russian VEGA balloons to Venus in 1985). The 4.6m balloon would have a polyester envelope, and be powered by a pair of ASRGs. The concept, however, was not pursued by ESA.

NASA's ongoing development of the efficient ASRGs prompted an aircraft concept called AVIATR. It would deploy into Titan's atmosphere and fly continuously for a year (never touching the ground) with two ASRGs providing power. Propulsion would be attained by a twin-bladed propeller driven by a rare-earth-magnet brushless DC motor. Science cameras mounted directly to the underbelly of the fuselage structure would be kept warm behind double-pane transparent windows. All of the flight control servo actuators would be internal to the fuselage and transmit their control forces and torques to aerodynamic surfaces via thermally insulating pushrods and torque tubes, with airflow baffles preventing excessive heat leaks. By flying at several metres per second (a good fraction of Titan's equatorial rotation speed), AVIATR could 'keep up with the Sun' and loiter over the day-side, thus maximising its communication windows with Earth to transmit prolific imaging data by a dish in its nose. The idea was developed to a level of detail enough for a preliminary costing. At $715 million it wouldn't quite fit as one of the Discovery missions NASA

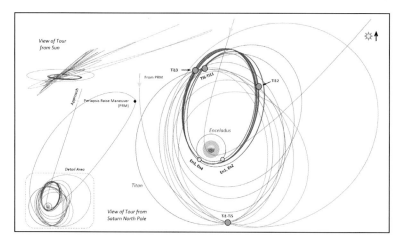

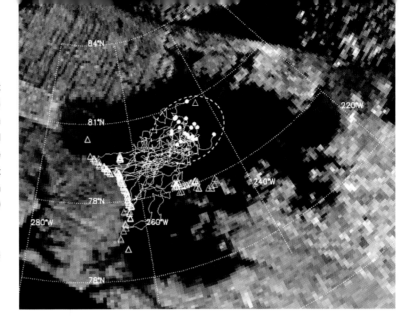

RIGHT Sailing on an extraterrestrial sea. A set of simulated TiME splashdown locations (dots) within an elliptical delivery footprint in Ligeia Mare near Titan's north pole. Winds from a global circulation model then make each virtual capsule drift in this trajectory calculation (lines) until it achieves landfall (triangles), typically 3-6 Titan days later. (Author)

was soliciting in 2010 but it could be a candidate for a New Frontiers mission.

One Discovery mission that was submitted was Journey to Enceladus and Titan (JET). This JPL concept would also use ASRGs for a more or less stripped-down TSSM mission making flybys of its two targets without entering into Titan orbit. Only an austere payload could be afforded: an infrared camera to map Titan more efficiently and at higher resolution than Cassini's VIMS, and a mass spectrometer to analyse Enceladus's plumes and Titan's atmosphere to higher molecular weights than could Cassini's INMS. The mission was not selected, however.

The only Saturnian mission to be selected for further study ($3M Phase-A) in Discovery in 2011 was Titan Mare Explorer (TiME), whose technical details were developed by Lockheed Martin and the Applied Physics Laboratory of Johns Hopkins University. A large standalone capsule (containing the fuel tanks and almost all the cruise systems, hence avoiding the requirement for a separate cruise stage) was to be launched to Titan and splash into Ligeia Mare, the second largest of Titan's seas. It would use a pair of ASRGs for electrical power and float nominally for at least six Titan days

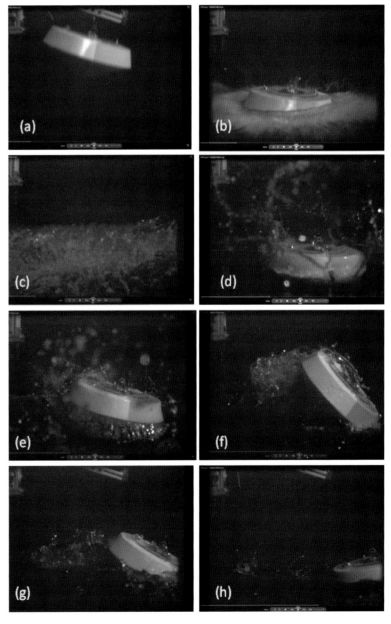

RIGHT High-speed video frames of a splashdown test of a 1/8th scale model of TiME at the Pennsylvania State Applied Research Laboratory. Note that although the capsule pancakes flat at impact, the resurge of the cavity in the water pushes the capsule over to one side (frame 'f'). In hindsight, this phenomenon may explain why many of the Apollo capsules capsized on splashdown, adopting an apex-down orientation until being righted by the inflation of airbags. (Author)

(three months) and possibly much longer. It would drift slowly, nudged by wind drag on its hull, camera mast and Medium-Gain Antenna.

Due to arrive in July 2023, TiME would exploit the fact that during the late northern summer on Titan, Earth barely sets below the horizon, enabling data to be sent directly to Earth without a relay spacecraft.

TiME would use a mass spectrometer to analyse the composition of the hydrocarbon sea, and a sonar would profile the depth. A meteorology package would study weather patterns and interactions such as evaporation and wave generation, while a camera studied clouds while 'at sea' and the shore when the capsule drifted within sight of land. Even with a modest payload, this idea was able to catch the public imagination and offer a credible mission with strong scientific appeal for a Discovery budget.

The TiME Phase-A study fleshed out the mission details, ranging from such obvious technical issues as the thermal design for operation in Titan's frigid but otherwise benign environment, to procurement and test schedules, to data archiving plans. The cryogenic ocean sampling system was prototyped and tested, wire insulation was tested for compatibility with liquid hydrocarbons, and even scale model splashdown tests were made to validate computer models of the loads and resurge dynamics. A major effort was the development of models for winds and the

statistics of waves: since the Applied Physics Laboratory does extensive US Navy work as well as NASA projects, experts were on hand to calculate sonar performance and develop simulations of the capsule motion in response to the wave field.

All this effort was for naught, however. In 2012, NASA's ASRG development faltered – although the Stirling engine was first developed by a Scottish clergyman in 1816, it proved difficult to maintain the ASRGs reliability at the same time as high performance, especially with the high internal temperatures and materials compatibility demanded by the radioisotope source. Without the ASRG, TiME could not fly. NASA instead selected a solar-powered Mars mission for its Discovery slot.

An imaginative concept that emerged a year or two later was a Titan submarine. This was not considered as a near-term prospect but as a 'blue sky' study for NASA's Innovative Advanced Concepts program. Using a (fictitious) Stirling power source, the submarine would cruise around Kraken Mare, Titan's largest sea, in 2040, when it was again summer in Titan's northern hemisphere. It would use a large electronically steered dorsal phased-array antenna to send data directly to Earth. An interesting challenge is buoyancy control – Titan's nitrogen dissolves in cold liquid methane much more than air does in water, so it is not practical to 'blow the tanks' in quite the same way.

Another Discovery opportunity opened in 2014, but because NASA did not offer RTGs or ASRGs this made it impossible to resubmit TiME (basically the last opportunity for it to fly before northern winter set in). A JPL team reworked the JET proposal to focus on Enceladus, a mission that was optimistically called Enceladus Life Finder (ELF). This would need colossal solar panels to provide enough power to send back even the modest amount of data from a mass spectrometer (the data-intensive imaging of the JET proposal had to be discarded) and the large panels might interact with Enceladus's plumes in ways that might be seen as risky. The concept was not pursued.

Cassini-Huygens answered many of the questions it was designed to address, but as always in planetary exploration, new questions were raised.

Of foremost interest are Titan's surface chemistry and methane hydrological cycle, and the plumes that emanate from the south polar region of Enceladus. At the time of writing (late 2016), there are no new Saturnian missions on the books. Nevertheless, ESA has issued a call for Medium-class proposals, and NASA intends to seek New Frontiers concepts for Titan and Enceladus.

There is no shortage of good ideas for future exploration at Saturn, especially of Titan where the vehicle possibilities seem boundless.

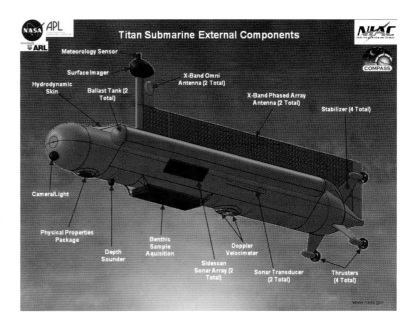

ABOVE The NASA Glenn Titan Submarine concept was a slender vehicle to minimise drag, with large buoyancy tanks to tolerate uncertainties in Titan's sea density. Four thrusters at the rear provide directional control and propulsive thrust and a streamlined camera mast and a large dorsal antenna would be above the 'waterline' when surfaced. *(NASA/GRC)*

Outer solar system exploration is not an enterprise that offers instant gratification, and it is not clear what scientific and engineering factors will win out in the choice of the next missions to this intriguing system. But as Cassini has shown, with patient persistence, adventure and discovery surely await!

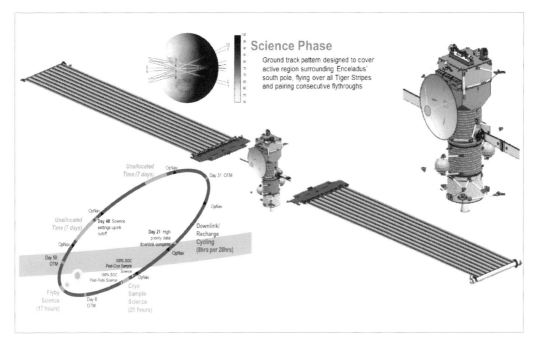

LEFT The Enceladus Life Finder mission would need huge deployable solar arrays to operate at Saturn. Its trajectory would fly through Enceladus's plumes several times to sample them using more sophisticated instruments than were available to Cassini. *(NASA/JPL)*

Further reading

Other volumes on Cassini include *Cassini at Saturn – Huygens Results* written by D. M. Harland published by Praxis-Springer (2007), and *The Cassini-Huygens Visit to Saturn: An Historic Mission to the Ringed Planet*, a NASA-commissioned history by the late M. Meltzer that was published by Praxis-Springer (2015). This latter book is particularly good on the political origins of Cassini.

The present author's involvements in Titan studies and the development and flight of Cassini-Huygens have been related in two books by R. Lorenz and J. Mitton, *Lifting Titan's Veil* published by Cambridge University Press (2002) and *Titan Unveiled* by Princeton University Press (2010).

Results from the prime mission are authoritatively given in the books *Titan from Cassini-Huygens* and *Saturn from Cassini-Huygens*, edited and written by Cassini mission scientists in 2009. Results from the Equinox and Solstice mission continue to be digested – the textbooks are still being rewritten!

Huygens at launch is described in detail in an ESA volume *Huygens: Science, Payload and Mission*, ESA Special Publication SP-1177. The design of, and expectations for the instruments of Cassini and Huygens, as well as overviews of the spacecraft, are described in a series of papers in *Space Science Reviews* in 2004–05.

The Huygens results appeared principally in *Nature* in December 2005 and in issues of *Planetary and Space Science* in 2007 and 2008 and *Icarus* in 2016. These and Cassini results papers can be found (but may need a university library subscription to access) with search tools such as http://scholar.google.com or the ADS database http://adsabs.harvard.edu/abstract_service.html.

Huygens data, including engineering housekeeping and in-flight tests, as well as detailed documentation, can be found on the ESA Planetary Science Archive http://www.cosmos.esa.int/ and (perhaps somewhat more conveniently) at the NASA PDS Atmospheres Node http://atmos.pds.nasa.gov/.

Cassini data are at various nodes of the PDS. Probably of most interest are the holdings of the Imaging Node. Again, in addition to the raw data there is copious instructive documentation. The User Guides of the individual experiment are particularly useful. The web forum unmannedspaceflight.com, and the blogs of the Planetary Society, have many helpful pointers on processing Cassini images.

The Cassini project web page is http://saturn.jpl.nasa.gov and it has links to images and mission timeline information. Some excellent CHARM presentations (Cassini Huygens And Results from Mission) can be found there too.

The Planetary Photojournal http://photojournal.jpl.nasa.gov/ has a good archive of the best images from Cassini-Huygens and other missions.

The design and operational experiences of the Cassini spacecraft's engineering systems are described in journal and conference papers published over the years by the AIAA and IEEE, and 'Cassini' keyword searches at the respective institutions' websites will help locate them.

Acknowledgements

Acknowledgement of image sources has been given, as required by publisher policy, and I am most grateful to my colleagues – too many to name – who dug around in drawers to find photographic material for this book (from a time when they were actually film photographs, rather than digital images), who have reviewed parts of the text, or dealt with the JPL bureaucracy in order to get material approved for ITAR (International Traffic in Arms Regulations) release. Even where colleagues have supplied material for which there was not space in this volume, I am appreciative nonetheless – seeing the whole picture has helped shape the story. Special acknowledgement must also be made to Dr Elizabeth Turtle, a Cassini imaging expert, who put up with her husband's distraction with this project.

Steve Rendle at Haynes helped shape the product and has been encouraging and patient, and David M. Harland has been not only a sharp-eyed editor of the final text, but was a valuable advocate for this project from its very beginning. Any errors of fact or attribution, however, are the responsibility of the author.

Abbreviations

A Amp (electrical)
AACS Attitude and Articulation Control Subsystem
AC Alternating Current
ACP Aerosol Collector and Pyrolyser
AGC Automatic Gain Control
ALC Automatic Level Control
amu Atomic mass unit
ANT Antenna Subsystem
ASIC Application-Specific Integrated Circuit
ASRG Advanced Stirling Radioisotope Generator
ATLO Assembly, Test and Launch Operations
AU Astronomical Unit

BCSS Back Cover Subsystem
BDR Battery Discharge Regulator
BITE Built-In Test Equipment
BIU Bus Interface Unit
BPSK Binary Phase-Shift Keying

C3 Trajectory energy parameter (twice the injection energy
    per unit mass = V2)
CAPS Cassini Plasma Spectrometer
CASA Construcciones Aeronáuticas SA (a Spanish
    aerospace company)
CASU Central Accelerometer Sensor Unit (on Huygens)
CCAFS Cape Canaveral Air Force Station
CCD Change Coupled Device (Visible Focal Plane Array)
CDA Cosmic Dust Analyser
CDS Command and Data Subsystem (orbiter)
CDMS Command and Data Management System (probe)
CDMU Command and Data Management Unit
CDS Command and Data Subsystem
CFRP Carbon Fibre Reinforced Plastic
CHEMS Charge-Energy-Mass Spectrometer (part of MIMI)
CIRS Composite Infrared Spectrometer
CMOS Complementary Metal Oxide Semiconductor
CNES Centre National d'Etudes Spatiales (France)
COSPAR Committee on Space
    Research (UN)
CRAF Comet Rendezvous-Asteroid Flyby

DA Dust Analyser (part of CDA)
DASA Deutsche Aerospace (a German aerospace
    company)
dB Decibel

DC Direct Current
DCSS Descent Control System
DCT Discrete Cosine Transformation
DDB Descent Data Broadcast
Delta-V Velocity Increment (change in spacecraft velocity)
DGB Disk Gap Band (parachute)
DISR Descent Imager/Spectral Radiometer
DM Descent Module (of Huygens)
DMA Direct Memory Access
DoD US Department of Defense
DoE US Department of Energy
DOR Differential One-way Ranging
DRAM Dynamic Random Access Memory
DSM Deep Space Manoeuvre
DSN Deep Space Network
DSP Digital Signal Processor
DST Deep Space Transponder (part of RFS)
DTTL Digital Transition Tracking Loop
DTWG Descent Trajectory Working Group
DVD Digital Versatile Disc
DWE Doppler Wind Experiment

EDAC Error Detection and Correction
EEPROM Electrically Erasable PROM
ELF Enceladus Life Finder
ELS Electron Spectrometer (part of CAPS)
EM Engineering Model
EM Equinox Mission
ENA Entry Assembly (of Huygens)
EPS Electronics Packaging Subsystem
EPSS Electrical Power Subsystem
EPT Experiment Polling Table
ERT Earth Received Time
ESA European Space Agency
ESOC ESA Space Operations Centre (Darmstadt,
    Germany)
EUV Extreme Ultraviolet Channel (part of UVIS)
eV Electron Volt

FET Field Effect Transistor
FFT Fast Fourier Transform
FGM Flux Gate Magnetometer
FIFO First-In/First-Out (data memory)
FM Flight Model
FPGA Field Programmable Gate Array

FRSS Front Shield Subsystem
FS Flight Spare
FSW Flight Software
FUV Far Ultraviolet Channel (part of UVIS)

GC Gas Chromatograph
GCMS Gas Chromatograph and Mass Spectrometer
GHz Gigahertz
GPHS General Purpose Heat Source
GPS Global Positioning System

HASI Huygens Atmospheric Structure Instrument
HDAC Hydrogen/Deuterium Absorption Cell
HGA High-Gain Antenna
HLC High Level Commands
HK Housekeeping
HOOD Hierarchical Object-Oriented Design
HRD High-Rate Detector (part of CDA)
HRG Hemispherical Resonator Gyro
HSP High Speed Photometer (on UVIS)
Hz Hertz (frequency)

IABG Industrie Anlagen Betriebs-Gesellschaft (a German industrial testing centre)
IBS Ion Beam Spectrometer (part of CAPS)
IMS Ion Mass Spectrometer (part of CAPS)
INCA Ion and Neutral Camera (part of MIMI)
INMS Ion and Neutral Mass Spectrometer
IRU Inertial Reference Unit
ISPM International Solar Polar Mission
ISS Imaging Science Subsystem
ISTS Inner Structure Subsystem
ITAR International Traffic in Arms Regulations (US)
IVP Inertial Vector Propagator

JET Journey to Enceladus and Titan
JPEG Joint Photographic Experts Group
JIMO Jupiter Icy Moons Orbiter
JPL Jet Propulsion Laboratory

K Kelvin, unit of absolute temperature
Ka-Band frequencies between 26–40GHz
KPT Kinematics Prediction Tool
KSC Kennedy Space Center
Ku-Band frequencies between 10–20GHz

LEM Lower Equipment Model
LEMMS Low Energy Magnetospheric Measurement Subsystem (part of MIMI)
LGA Low-Gain Antenna
LHCP Left Handed Circularly Polarised (radio wave)

LNA Low Noise Amplifier
LV Latch Valve

MAG Dual Technique Magnetometer
MBB Messerschmitt-Bölkow-Blohm (a German aerospace company)
MCP Microchannel Plate
MEA Main Engine Assembly
MHW Multi-Hundred Watt
MHz frequencies in millions of cycles/sec
MIL-STD-1553 A US Military Standard serial data bus specification
MIL-STD-1750A US Military Standard computer instruction set architecture
MIMI Magnetospheric Imaging Instrument
MLI Multi-Layer Insulation
MMH Monomethyl hydrazine
MMRTG Multi-Mission Radioisotope Thermoelectric Generator
MS Mass Spectrometer
MTT Mission Timeline Tables
MTU Mission Timer Unit (on Huygens)

N Newton (a force about equal to the weight of an apple on Earth)
NAC Narrow-Angle Camera (part of ISS)
NASA National Aeronautics and Space Administration
NSI NASA Standard Initiator
NTO Nitrogen Tetroxide – Bipropellant Oxidiser

ODM Orbit Deflection Manoeuvre
OPNAV Optical Navigation
ORS Optical Remote Sensing

PCDU Power Conditioning and Distribution Unit
PDD Parachute Deployment Device
PDRS Probe Data Relay Subsystem
PDS Planetary Data System
PDT Pointing Design Tool
PHSF Payload Handling and Servicing Facility
PJM Parachute Jettison Mechanism
PLL Phase-Locked Loop
PM Prime Mission
PMS Propulsion Module Subsystem
POSW Probe On-Board Software
PPS Power and Pyrotechnics Subsystem
PROM Programmable Read-Only-Memory
PSA Probe Support Avionics
PSE Probe Support Equipment
psi Pounds per square inch
PSS Probe Support Subsystem

PST Polling Sequence Table
PTT Probe Transmitting Terminal
Pu-238 Plutonium-238
PVDF Polyvinylidene flouride
PWA Permittivity and Wave Analyser (part of HASI)
PYRO Pyro Unit
PZT Lead zirconate titanate (a piezeoelectric ceramic)

RADAR Cassini Radar Subsystem (RAdio Detection And Ranging)
RAM Random Access Memory
RASC Revolutionary AeroSpace Concepts
RASU Radial Accelerometer Unit (on Huygens)
RAU Radar Altimeter Unit (on Huygens)
RBOT Reaction Wheel Assembly Bias Optimisation Tool
RCS Reaction Control Subsystem
REU Remote Engineering Unit
RF Radio Frequency
RFE Receiver Front End
RFS Radio Frequency Subsystem
RFIS Radio Frequency Instrument Subsystem
RHCP Right Handed Circularly Polarised
RHU Radioisotope Heater Unit
R_J Jupiter radii
ROM Read Only Memory
RPM Revolutions per minute
RPWS Radio and Plasma Wave Science
RS Radio Science
R_S Saturn radius
RTG Radioisotope Thermoelectric Generator
RTI Real Time Interrupt
RUSO Receiver Ultra Stable Oscillator
RWA Reaction Wheel Assembly

SAEF Spacecraft Assembly and Encapsulation Facility (at KSC)
SAR Synthetic Aperture Radar (a RADAR imaging mode)
SASW Support Avionics Software
S-Band radio frequencies between 2–4GHz
SCAS Science Alignment System
sccm standard (i.e. room temperature/pressure) cubic centimetres per minute
SCET Spacecraft Event Time
SED Spin Eject Device
SEPS Separation Subsystem
SEU Single Event Upset
SIC Support Interface Circuitry
SID Star Identification
SIS Software Interface Specification
SM Solstice Mission
SM2 Special Model 2

SOI Saturn Orbit Insertion
SO2P Saturn Orbiter/Dual Probe
SOPC Science Operations and Planning Computer
SRMU Solid Rocket Motor Upgrade
SRU Stellar Reference Unit
SSA Sun Sensor Assembly
SSP Surface Science Package
SSPS Solid State Power Switch
SSR Solid State Recorder
SSRS Solid State Recorder Subsystem
STPM Structural, Thermal and Pyro Model
STRU Structure Subsystem

TAE Titan Aerial Explorer
TAT Time-Altitude Table
TCM Trajectory Correction Manoeuvre
TCXO Temperature Controlled Crystal Oscillator
TEMP Temperature Control Subsystem
THSS Thermal Control Subsystem
TiME Titan Mare Explorer
TOST Titan Orbiter Science Team
TPS Thermal Protection System
TSSM Titan Saturn System Mission
TTL Transistor-Transistor Logic
TUSO Transmitter Ultra Stable Oscillator
TWTA Travelling Wave Tube Amplifier

UAV Unmanned Aerial Vehicle
UDMH Unsymmetrical dimethyl hydrazine
UEM Upper Equipment Module
USAF US Air Force
USM Umbilical Separator Mechanism
USO Ultra Stable Oscillator
UT/UTC (Coordinated) Universal Time
UVIS Ultraviolet Imaging Spectrometer

VCXO Voltage Controlled Crystal Oscillator
VHM Vector Helium Magnetometer
VIMS Visual and Infrared Mapping Spectrometer Subsystem
V Volt (electrical)
VLA Very Large Array (radio telescope)
V/SHM Vector/Scalar Helium Magnetometer

W Watt (electrical)
WAC Wide-Angle Camera (part of ISS)
WWW World Wide Web

X-Band radio frequencies between 8–12GHz
XM Extended Mission (also Equinox Mission)
XXM Extended-Extended Mission (also Solstice Mission)

Index